Praise for **THE NEUROTOURIST**

"A fascinating exploration of the [illegible] s so far… her hesitant participa[illegible] brutally honest descriptions of the experts, add a welcome dose of humour."

New Scientist

"Riveting. Lone Frank has selected the most intriguing issues currently engaging scientists and philosophers, and presented them in a way that will engage anyone who posses the organ she writes about."

Rita Carter, author of *Mapping the Mind* and *Multiplicity*

"Does a great job of exploring the impending neurorevolution and the sometimes scary consequences of neuro-technology."

Susan Blackmore, ***BBC Focus***

"Think that you know yourself. Think again. The coming neurorevolution will destroy your certainties – but may set you free. Arm yourself. Read this book."

Armand Leroi, author of *Mutants* and
Professor of Evolutionary Developmental
Biology at Imperial College London

THE NEUROTOURIST

POSTCARDS FROM THE EDGE OF BRAIN SCIENCE

LONE FRANK

ONEWORLD
OXFORD

A Oneworld Paperback Book

Previously published in Danish as *Den Femte Revolution*,
Gyldendal, 2007
First published in English as *Mindfield* by Oneworld Publications, 2009
Reprinted, 2009
This edition published 2011
Reprinted, 2012

A CIP record for this title is available
from the British Library

ISBN 978–1–85168–796–1

Cover design by Leo Nickolls
Printed and bound in Great Britain by
TJ International Ltd, Padstow, Cornwall

Oneworld Publications
185 Banbury Road
Oxford OX2 7AR
England

To my father

My "perfectly neutral and completely unbiased reader"

Contents

Preface

This is not the definitive book about the brain – a book that exhaustively describes everything we know about the 1300 grams of tissue that resides between our ears. *That* book does not exist and probably never will. This book is a snapshot of a particular moment, highlighting a number of cutting-edge developments in brain research, and providing a glimpse into some of the key personalities involved.

Many years ago, as a newly-trained biologist, I dabbled in brain research and, even though I chose a different path, its deep fascination has never waned. But fascination is not a goal in and of itself. Brain research is not just exciting, interesting and entertaining for those of us who aren't in the field – it is literally shaping how we think and how our society is going to develop. This all-encompassing significance is getting far too little attention, and I hope this book will contribute to a shift in focus.

Mindfield could only be realized because of the enthusiastic people who have supported the project. I owe a tremendous debt to H. Lundbeck A/S and its managing director Claus Bræstrup. I am also grateful to the Literature Committee of the Danish Arts Council and to the Ulla and Mogens Folmer Andersen Foundation for generously supporting both my travels and the translation of the manuscript. Finally, a great thanks to my agent, Peter Tallack, for believing this could be done.

Lone Frank

1

BRAINY REVOLUTION

It's an awkward situation. Tears are running down my face, and I'm quite sure the good-natured man on my right has noticed them, as he speaks to me. I try to blink them away, opening and closing my eyelids again and again without any apparent effect against the formaldehyde fumes wafting up from the white plastic bucket in front of me. I'm holding a human brain – half a human brain, to be exact – whilst trying to concentrate on what the man is telling me, as he gesticulates and explicates, clearly expecting some sort of reaction from me. The brain resting in my right hand is split lengthwise, revealing its knurled structures and inner cavities. There is something undignified about the way its halved cerebellum dangles over my wrist.

"You say you want to write a book about the brain. Then, I suppose a good place to start is to look at it up close. How's your anatomy? We'll start with the easy stuff: this thick white band is the *corpus callosum*. Its 200,000 transverse nerve fibers allow the two hemispheres to communicate. I call it the Brooklyn Bridge of the brain."

It's not the first time George Tejada has used that line, and it's not the first brain he's studied at close quarters. George is head technician at the Harvard Brain Tissue Resource Center, the world's largest brain bank, and he personally handles every one of the three hundred-odd human brains that are donated each year for research and end up here in the Mailman Building at McLean Hospital in Belmont outside Boston. George is a slender, middle-aged man, dressed in basic, green hospital scrubs. His graying buzz cut and concise movements give him an air of efficiency and enterprise, but his Spanish accent softens the image a bit. All in all, a man you can trust with your donated organs.

"This is the hippocampus." George reaches out and traces a curvature along the underside with his little finger. "This is what stores all your experiences. Without the hippocampus you're nothing."

Of course, in theory, I'm perfectly familiar with the function of this twisting, sausage-like structure, but I've never seen it before in real life. I move my face closer to the brain. I'm not really able to distinguish anything in particular in the beige mass. Actually, all I can think of is how much brain tissue reminds me of pickled mushrooms. The strong fumes make me shed a tear on the brain stem. George doesn't see it or pretends not to. He simply turns the brain over and asks me to note how the folds on the outside of the cerebral cortex are much less full than they should be. Instead of being filled out and having an almost smooth surface, it has deep hollows that switch back and forth like a dried walnut.

"Severe atrophy."

"Alzheimer's?"

George nods, and I feel like an A student. The merciless progression of dementia has dissolved precious tissue and left a shrunken, compromised organ. From having been a well-functioning person – "Woman, the brain belonged to an elderly woman" – she moved deeper and deeper into a darkness without memories, without language, and finally without consciousness. Because of her illness, she ended up here in a white plastic pail with no identity other than number B6782. The Brain Bank collects diseased brains in order to send tissue samples to researchers throughout the world. *From knowledge will come a cure*: this slogan appears on the front page of all the Brain Bank's pamphlets. Researchers study changes in the tissue to understand what is going on in conditions such as Alzheimer's, Parkinson's, schizophrenia and bipolar disease.

"I can't give you the exact amount off the top of my head, but it costs thousands of dollars to get a single brain through the standard procedures," says George. And there is a professional pride in his meticulous explanation of how this sensitive organ must be removed from its former owner immediately after death and deposited with the Brain Bank within twenty-four hours.

"We always have somebody on call, always."

Please contact the Brain Bank before the brain is removed, counsels the protocol sent out to nursing homes and hospital wards attending to a donor. Do not embalm the deceased before the brain is removed and please place the deceased in a refrigerated environment as soon as possible and no later than six hours after death has occurred. The Brain Bank will send you the necessary shipping materials.

When the chilled brains arrive at McLean, they are immediately cut into two halves. In the laboratory, there are pictures of young Lou, with a Plexiglas screen in front of his face and a big smile for the photographer, slicing through yet another bloody brain. "Notice how he moves the knife from below and upward," George remarks. "That makes for a beautiful brain stem."

One half goes directly into the freezer and the other is preserved in formaldehyde, suspended in the thick liquid in plastic buckets, until it is taken out and sliced into cross-sections.

They are then dispatched as tissue samples to researchers who have applied to and been approved by the Bank. The brains also pass through the hands of trained pathologists, who make sure that the right diagnosis was made, and the cross-sections undergo staining and characterization. Usually, one brain arrives every day and, today, there are almost seven thousand in the repository. The Bank has been in existence since 1978 and is still contributing new knowledge. In the past, the Bank provided tissue samples that helped identify the genetic defects behind Huntington's chorea, an incurable degenerative disease resulting from the death of certain brain cells. More recently, the director of the Brain Bank, Francine Benes, has been focusing research on schizophrenia and bipolar disease. By studying brains donated to McLean, she has ruled out a hypothesis that the two diseases have something to do with degeneration and cell death, which indicates that they may rather be associated with defective connections in the brain.

"I don't do research myself," says George, Yet, in his six years at the Bank his dealings with dead tissue have never become routine or everyday. He makes a sweeping gesture with both arms.

"I love my job. I never get tired of talking about it. The brain is a deeply fascinating subject for people, and it still has a powerful effect on me. You can't help but be moved knowing that this is a person, this thing you have in your hand was a human being."

He's right. It's very difficult not to be moved. Of course, a pair of rubber gloves separates me and the deceased woman's right hemisphere, but I almost feel a tingling as if a current were running through the 700 grams of cold tissue. It's a strange sensation, a tremulous, unsteadying and actually quite unpleasant sensation. Entirely unexpected.

"You're a biologist!" I tell myself. "You've dissected everything from earthworms to rabbits without a peep. You carved up rats for years to cultivate their brain cells without a tremor. At any rate, you didn't feel anything in particular."

But now all my cool academic interest is gone. Standing here with the remains of B6782 almost makes me want to cry – tears that are not due to the formaldehyde. Thin, cold needles prick the flesh up and down my back, and the uneasiness releases little balls of lightning in the pit of my stomach. It is just as George says: I'm holding the very essence of a person in my hands. This massive blob was – merely a week ago and for an entire life – the innermost core of another human being. All the thoughts, feelings and unconscious desires of this person were electrical impulses ceaselessly leaping between individual cells along a delicately branching network of axons and dendrites. Fingering this tissue somehow feels like a transgression. The moment is horribly intimate.

At the same time, the surroundings and the circumstances are astonishingly mundane. The room we're in is tiled in gray linoleum, illuminated by white neon lights, spotless and anonymous – reminiscent of a veterinarian's clinic after the clients have all gone home. A set of steel scales hangs from the ceiling, like at a butcher's shop. There is a row of glass-encased cabinets along the walls and a fountain pen lying parallel to a yellow notepad on one of the desks. Everything is practical and purposeful, without ornamentation. This is a workplace. And George is a man who brings your mind back to the concrete.

"Look. This is what happens when you eat too much junk food."

He pulls the stub of an artery away from the underside of the exposed brain. It is bright yellow and doesn't tally with the pale, diluted color scheme of the rest of the brain.

"Feel how hard it is."

I dutifully squeeze the thick artery with two fingers and feel its hardness. Like plastic. George suddenly turns, walks to the corner and gets yet another receptacle, which opens with a snapping sound. He quickly puts his hands down in the liquid and brings up another hemisphere. He holds it next to mine; you can see that it is larger and its form fuller.

"See, that's how a control brain is supposed to look."

A control brain. That is, an ostensibly normal organ like the one George and I still carry around in our skulls. I put the disease-ravaged B6782 back into the viscous fluid and feel like I'm putting down a burden. George looks as if he wants to say something but simply lets a smile play on his lips.

"Apparently, the world can't get enough of the brain." The remark comes from the door. "We always have somebody visiting. You've come from Denmark. Next Tuesday, a team of researchers is coming from the Karolinska Institute in Stockholm, and the week after that we're getting a visit from German TV."

Timothy Wheelock extends his hand, and I peel off my wet glove. Dr Wheelock, as he is called here, is sporting a canary yellow shirt and looks like a clone of Bill Clinton during his happier days in the White House. Wheelock is the head of histopathology at the brain bank. He is the one in charge of cutting micrometer-thin slices from particularly important parts of the incoming brains, so they can be studied and a precise characterization and diagnosis made. Wheelock happily joins our excursion.

"People come in large tour groups. Of course, there are media people running around all the time, but there are also teams of nursing students and an endless procession of high school classes and librarians."

"Librarians?"

"Yes, we've had quite a few of them. Don't ask me why. But the high school students are my favorite. These kids are crazy about looking at all the stuff we've got. They think it's cool and a little creepy at the same time. They just lap it up, when I lay out all my stained samples, and the best thing is going to the storage room."

The storage room is like a modern vision of eternity. From floor to ceiling, there are brains in slices and smaller fragments collected over the years and stored in transparent tupperware containers.

"A brain donation is an invaluable gift to neuroscience research," explains the attractive informational brochure. And a glossy, confidence-inspiring light-brown folder deals with the religious aspects of a *post mortem* donation.

> Many people find this decision difficult and complicated. It is a decision that makes many people examine their innermost thoughts about death – whether there is life after death and what makes up the soul.

Fortunately, if you are Protestant, Catholic, Greek Orthodox, Muslim, Jewish or Buddhist, you can find religious support for donating your brain to science. Pope Pius XII was an early standard bearer for tissue donation, and his successors hold the banner high. The Orthodox Rabbi Moses Tendler goes so far as to say that organ donation is actually a duty under certain circumstances.

Each psychiatric and degenerative disease has its own special brochure. There is also one for normal control brains. In the back, you can find a form on which you can register your preliminary interest. "If you are interested in making a brain donation, we recommend the following steps: 1) Discuss it with your family and inform your doctor. 2) Fill out and return the attached questionnaire."

Despite the fact that postage is even paid in advance, donors are hard to come by. Healthy folk, who would function as a control group, are especially reluctant to be parted from their brains. The sick are more willing, possibly because the bank has good connections with patient associations, which provide information and encouragement. Most control brains come from the spouses of sick donors. Naturally, I imagine that the employees at the brain bank pledged their organs long ago; but both George and Wheelock remain silent, when I ask. The two gentlemen smile wryly, look at each other and then down at the table. They have never really considered it, they say, and Wheelock's voice is strangely faltering.

"If I got Parkinson's, it probably wouldn't feel so strange, but there's nothing wrong with me. What about you, George?"

"I would leave it up to my family. They're the ones who'll have to live with the decision."

The brain bank does not accept donations from abroad – the transport time is too long. But if they took Danish brains, would I give them mine? I already carry an ordinary donor card in my purse, so my serviceable spare parts can be used if I meet with an accident. But it feels different with the brain. To think that Timothy Wheelock would inspect its most minute details and George Tejada would cut it into small pieces, pickle it in a jar, and keep it in a storage room. Or even worse: That some callous journalist would fondle it and describe the experience in some tawdry publication.

Take my liver, my kidneys, my heart – fine, they're just organs. But my brain – that's me! I can almost endorse Sherlock Holmes' adage: "I am a brain, my dear Watson, and the rest of me is a mere appendage."

But that's the way it is – it's sinking in that we, each of us, *are* our brains. Not so very many years ago, there was fierce opposition to heart transplants, because the heart was somehow associated with the self. Today, we all know – and feel – that the heart is simply a muscle, a pump that can be replaced, like the carburetor in a car. As the heart decreased in importance, the soul has ever so gradually become equated with the brain.

"Know thyself", it said above the entrance to the oracle at Delphi, and more than two thousand years later, we're still on the same quest. "Who am I and what does it mean to be human?" we ask. But we are asking in a new way. Whereas, before, the speculations turned to culture and the psyche, which was strangely disconnected from the organism, the physical brain is now prominent and steadily becoming the reservoir and end station for all the questions we ask about human nature and existence.

What goes on in the brain, when we love or hate ? What areas of the brain are active when people gamble or hunger for alcohol or cocaine? What's wrong with the brain of a violent criminal? Where do emotions reside, and how are thoughts generated?

Neuroscience is the new philosophy, some say, and there is no doubt that brain research is the hottest topic a scientist can dabble in, and the most distinguished thing you can put on your calling card. A while ago I heard the American philosopher Daniel Dennett explain why this is the case. He was in transit on his way from one interview to the next but agreed to meet me at Boston airport, where he generously provided a meal of oysters and lots of white wine. The interview focused on his latest book, *Breaking the Spell*, in which Dennett argues that religion is a natural phenomenon. There are no gods, just stubborn, irrational ideas that only exist between our ears. Ideas born from crackling electronic signals between brain cells that are carried on by language and upbringing from one generation to the next. At one point in an extended exposition, Dennett suddenly stopped to utter something that sounded like a prophecy.

"The next generation of geniuses will appear in brain research. Once it was particle physics that attracted the brightest young people, then it was DNA and genome research, but now it's the neurosciences. Because this is where you can answer the big questions."

With his white hair and beard, Dennett had taken on the air of a Biblical patriarch; looking serious, he pointed at me with his oyster fork. I tried to respond with something clever, but since he'd been so kind as to refill my wine glass several times, I could only produce a meek affirmative remark. Not long afterwards, I happened to think of our conversation when I stumbled upon an excerpt from author Tom Wolfe's collection of essays, *Hooking Up*. The book was published in 2000, in the heyday of information technology, but even then Wolfe could already glimpse the horizon beyond the digital landscape. "If I were a student today, I don't think I could resist going into neuroscience," he writes. "Here we have the two most fascinating riddles of the twenty-first century: the riddle of the human mind and the riddle of what happens to the human mind when it comes to know itself absolutely."

And now the students are flooding in. The most prestigious universities from Harvard and MIT to Princeton and Cambridge have all established expensive neurocenters equipped with high-tech machinery and special grants. Evidence for how hot the field of neuroscience is can be seen in the fact that these centers are where wealthy patrons choose to place their university sponsorships and have their names above the door. These centers have head-hunted the greatest talents in the field and are able to examine the brain at all levels from mapping genes to uncovering signal molecules to neuropsychological studies of the entire person.

One particularly interesting development is the migration to neuroscience from other fields. Not only psychologists but also sociologists, anthropologists, researchers of religion, and philosophers. A good example of the latter is the American philosopher Sam Harris, who left the ivory tower, exchanged his quill pen for an MRI scanner and aimed for a Ph.D. in neuroscience. Why the radical gear shift, I asked when I met him recently, and the answer came quickly and without a hint of uncertainty.

"Originally, I planned to do a Ph.D. in philosophy and specialize in the philosophy of mind. But I got so tired of listening to philosophers talking in circles about the brain. It was blindingly clear that, if I wanted to know more about the human mind – about consciousness, rationality, faith and other aspects of our subjective self – well, then I had to learn more about the brain."

Today, he is hard at work expanding our common knowledge. Harris uses his scanner to investigate how concepts like "believing in something," "denial of something" and "doubt about something" are actually manifested in the brain. He is mapping the relevant networks and circuits of brain cells to create an understanding of the processes involved when we piece together our picture of reality.

A generation ago, this sort of question was inconceivable, outside the reach of natural science. Neuroscience was about explaining the anatomy of the brain and studying the complicated biochemistry and details of individual nerve cells and how electronic signals move between cells. Researchers sat bent over Petri dishes and test tubes or fiddled with small pieces of detached tissue. Neuroscience was, in reality, a branch of physiology. Thanks to new technology, it is about to be transformed into the queen of sciences.

The miracle of new imaging technologies – PET, MRI, SPECT – is that they provide an opportunity to look directly into the living, working brain. With scanners, we have a peephole or perhaps even a panoramic window into the thinking and feeling universe between our ears. Imaging technology opens up the possibility for comparing activity in the brain directly with actions, sensations and choices – indeed, even things as abstract as thought, emotions and attitudes. Here we have a true mirror of the soul.

Even if the information flooding out of labs and research centers can sometimes seem rather esoteric and specialist, the knowledge it brings is pertinent for rest of us. Because we find ourselves on the threshold of a neuroscience revolution. Or as neurologist Vilyanur Ramachandran of the University of California at San Diego has called it: the fifth revolution – the latest in the series of scientific great leaps forward that have turned our worldview upside down and caused great intellectual and social upheavals.

When Copernicus shook things up in the sixteenth century by yanking the Earth from its place at the center of God's universe, the church was incensed and scientists risked being burnt at the stake. In the mid-1800s Charles Darwin caused upheaval by taking man himself off his pedestal. With evolution and natural selection the human race was no longer a specially favored and *created* creature but a mere descendant of primitive primordial forms. A bit later Freud revolutionized the view of the human mind by introducing the unconscious and puncturing the entrenched assumption that we have total control over ourselves. And today, our world view is still in the process of adapting to the discovery of DNA. In the age of genetic engineering life itself has lost its special status and living beings become like any other malleable material. And many people are disturbed by this, as we can see in the public protests – from the passionate resistance to Frankenfood to the running battle over stem cell research.

What sort of a shake-up are we facing with the neurorevolution? This is where Tom Wolfe puts his finger on the most profound and central question of our time: what happens when the human mind comes to know itself completely?

What *will* all this come to mean? How is the fifth revolution going to influence our vision of what it is to be human – of what it is to be a self? And what consequences will it have? Will it change our personal lives? Will it even lead to fundamental changes in the social order?

Brain researchers dissect everything that makes us human and anchor all sorts of phenomena we have been accustomed to considering incorporeal in soggy cell structures – and, thus, ultimately in the exchange of chemicals, in electrical signals, in processes that slavishly follow the basic laws of physics. Religion is gone, pigeonholed as a prosaic neurological phenomenon; moral choice is no longer an expression of spiritual development or integrity but is ascribed to automatic processes that are planted in us all by a blind, value-neutral evolution. And then there is the modern-day quest for the Holy Grail: to explain consciousness itself, the foundation for the individual's subjective experience of the world.

We are sliding towards what you might call neurocentrism, where the very essence of what it is to be human is located in the brain, and what is in your brain determines who you are.

This is in contrast to the DNA centrism that has flourished over the past few decades. Here, the focus has been on genetic material and a belief that the genetic code is a sort of key to the essence and potential of the individual person. It is the Genes 'R' Us world view. A trend that culminated at the turn of the millennium with the conclusion of the human genome project. We saw Clinton and Blair proclaiming on live TV that "the book of life" had now been deciphered and the road was open to understanding the relationship between genes and organisms. In public discourse people talked about genes "for" this and genes "for" that. Everything from the need to smoke to an unfortunate predilection for adulterous men was put down to genes. And there wasn't much you could do about it, because the DNA combination you drew in the genetic lottery cannot be changed.

Neurocentrism will be critical for our self-understanding. Especially because it is so obvious that the brain is a tremendously dynamic system. There is simply not enough information in just under 30,000 genes for them alone to determine the pattern of the brain's network and communication links. And the three pounds of tightly-packed cells constitute an organ under constant, lifelong change. Cells are continuously refurbished, and life's abrasive stream of impressions causes old communication links to break down and new ones to be established. It is almost inconceivable, but every second something in the range of a million new links are formed. Simultaneously, more subtle processes are at work that strengthen or weaken existing links, prioritizing or de-prioritizing information. Finally, there are cells that die and cells that are born. A deeper understanding of this constant flux and especially an understanding of how it all connects with our inner lives and external behavior also opens up possibilities for modulating the processes of the brain. And thereby, in the final instance, for molding the self – the core that for each of us is "me". In that way neurocentrism represents a big step away from a deterministic view of who we are and what our lives can be.

If neuroscience is the new philosophy, this must also make neuroscientists the philosophers of our time. They are the ones who can see where the laboratories are going, and they can point out which technologies may emerge from the experiments. They are also uniquely positioned to see the challenges.

To dig deeper into how neuroscientists see the future I have brought together an elite group of researchers who are working to gain insight into areas of existential significance. People who, in their study of the brain's secrets, have put a question mark on such fundamental phenomena as religion, faith and morality. People who are shaking up our views of reason and emotion by revealing how rationality and unconscious automatism fight battles inside us and how the outcome determines our view of the world and our everyday choices. People who are unraveling how one of our most definitive and amazing abilities – empathy – is created in a small, dispersed group of neurons.

While researchers undertake their academic exercises, the conversion of scientific knowledge into business enterprises is taking place elsewhere. The advertising world is working on how to market commodities directly to our receptive nervous system and eager entrepreneurs are trying to market brain-scanning technologies as the infallible lie detectors of the future.

All this raises questions and scientists have begun in their small way to speak out. In 2004, a team of highly esteemed brain researchers headed by Nobel prize winner Eric Kandel banded together to warn against a possible future in which healthy people are forced to take drugs to increase or optimize normal brain functions. Such doping, they believe, may be the end result of putting the brain center stage. In his book *The Ethical Brain*, respected neuroscientist Michael Gazzaniga says that knowledge about the brain must lead to nothing less than a universal ethics. And as he concludes, "Our species wants to believe in something, some natural order, and it is the job of modern science to help figure out how that order should be characterized."

There is a growing awareness that neuroscience is no longer just about understanding and curing brain diseases but is on its way to having far more sweeping effects. As editors Jean Decety and Julian Paul Keenan put it, when the journal *Social Neuroscience* hit the streets in 2006: "As social neuroscience develops, it will certainly challenge our ways of thinking about responsibility and blame, and have an impact on social policies."

It is time to bring the discussion out of the specialists' domain, and to acknowledge that we are all tumbling headlong into the age of the brain.

Introducing the Brain

The ultracondensed overview

"The most complex object in the universe" weighs about 1300 grams, and contains a hundred billion brain cells – neurons – that are all connected by hundreds of trillions of communication links. Its general anatomical organization reflects the process of evolution. You can look at the human brain as a sort of Lego kit in which ever more advanced structures have been laid down on top of each other over the course of evolutionary history.

At the base are the most primitive parts – the cerebellum and the brain stem. They control, respectively, basic movement and heart and lung functions. Together, they correspond roughly to the reptile brain. In the evolution from reptile to mammal, other additional structures evolved, which are today packed around the brain stem. Innermost are the basal ganglia, which help regulate and modulate movement; then comes the limbic system, which is often characterized as the foundation of our emotional life. This system consists of specialized structures such as the cingulate gyrus, the hippocampus and the amygdala, as well as the thalamus and hypothalamus. The latter two receive information from the rest of the central nervous system and, through hormones, regulate basic drives such as hunger, thirst, sleep and sexual drive. The hippocampus does a number of jobs involving memory, while the amygdala is involved with our emotional repertoire in various ways.

On the outside of these tightly-packed structures is the cerebral cortex – the characteristically coiled surface – and it is this that is particularly distinctive in humans. The cortex makes up eighty percent of our brain's overall mass; in rats, for example, it constitutes only thirty percent. Roughly speaking, the cerebral cortex is divided into four lobes – the occipital lobe in the back, the parietal lobe on top of the head, the temporal lobe around the temples and ears, and the frontal lobe up front. The division reflects a certain division of labor and specialization: the occipital lobes deal with the sense of sight, while the other senses are processed in the parietal and temporal lobes, which also process language. Much of what we call higher cognition takes place in the frontal lobes, that is, processes that have to do with conscious thinking, understanding and planning. It is here in the "CEO of the brain," as the frontal lobes are sometimes called, that conscious decisions are formed, and any action that has an element of choice emanates from here.

The basal ganglia and cerebral cortex are divided into two halves or hemispheres, which mirror each other anatomically. Pretty much every region has a left and a right version, and the two brain halves are engaged in intense communication through three massive nerve bundles, the largest of which is the corpus callosum. As for movement, each hemisphere controls the opposite side of the body, and there is also a certain hemispheric specialization with respect to mental functions. For example, grammatical language processing and linear mathematical reasoning take place primarily in the left hemisphere, while more abstract mathematics, spatial manipulation and language functions such as intonation, primarily take place in the right hemisphere.

The knowledge we have about what areas of the brain take care of what types of tasks comes from countless studies over time. Studies of people with specific brain injuries and experiments on animals have indicated a number of functional areas, particularly in the cerebral cortex, and have mapped connections between them. Nevertheless, our understanding of how the brain functions is still very rudimentary. With modern scanning techniques and experiments on the living brain, there has been an explosion in the ways we can gain deeper insight into how these functions arise through the interplay of systems of neurons and areas of the brain. Researchers are facing a gigantic task, and the work has only just begun.

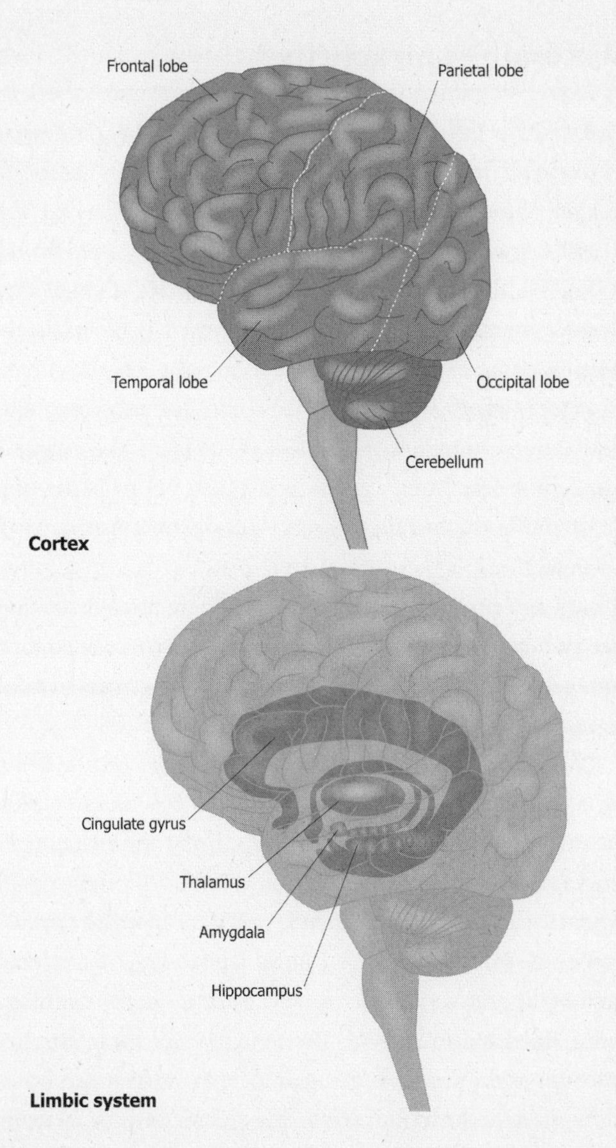
Frontal lobe
Parietal lobe
Temporal lobe
Occipital lobe
Cerebellum
Cortex
Cingulate gyrus
Thalamus
Amygdala
Hippocampus
Limbic system

The brain

Measuring brain activity

The various technologies for quantifying brain activity use either indirect imaging techniques or direct electrical measurements. At present, the key technologies are:

PET (*Positron emission tomography*) A scanning method that measures radiation from injected radioactive materials. Can, for example, measure the brain's use of glucose as an indirect measure of activity.

fMRI (*Functional magnetic resonance imaging*) Measures, via radio waves, the blood flow and thereby activity in different areas of the brain.

SPECT (*Single photon emission computed tomography*) Measures blood flow and activity by way of gamma rays emitted from a radiotracer.

EEG (*Electroencephalography*) Measures electrical charges from major groups of brain cells or areas of the brain using electrodes outside the cranium.

2

Finding God in the Synapses: Your Own Personal Jesus

I have never understood religion. Even as a child I was baffled by the fact that so many people believed in the existence of something for which I could not see the slightest evidence.

When I was young I would ask "Have you ever seen God?" And otherwise perfectly rational adults would reply "No, but I *believe* in him." "Have you ever spoken to him?" "No, but I know he exists." "OK, so he must answer when you pray to him for something?" "No, it's not that simple, sweetie" – and here there was often a little pat on the head – "God is not a vending machine. He has a greater plan for us about which we know nothing. That's just the way it is. We have to accept it."

Yes, well. That's hard for the average six-year-old intellect to digest. Particularly as the same grown-ups clearly didn't believe in Santa Claus or the Tooth Fairy. And I had already worked out that when those two appeared to grant my wishes in the form of presents or coins under my pillow there was a much more mundane hand at work.

"Denmark's most fervent atheist," my father calls me these days, when we touch on the subject of religion; less good-natured interlocutors have used the label "scientific fundamentalist." Now I'm on a pilgrimage worthy of a fervent atheist. I've journeyed to the distant and not particularly charming Canadian mining town of Sudbury to have a religious experience produced with the help of modern neuroscience. I have an appointment with Professor Michael Persinger, who is affiliated with the local Laurentian University. The professor has developed a technique to stimulate certain parts of the brain and thereby create a sort of mystical experience that some people have likened to a religious revelation or an encounter with higher powers.

His unique contraption has been dubbed the *God Helmet*. It looks like a souped-up version of an ordinary yellow crash helmet. On the inside, it has been equipped with magnetic coils, which Persinger and his people program to emit a complicated pattern of weak magnetic pulses directed at strategically selected areas of the cerebral cortex. Over the last two decades, Persinger has had more than a thousand people participate in his experiments with the special helmet, and almost eight out of ten report what the professor calls a "sensed presence."

The research subjects have a clear and undeniable sensation of being in the presence of "someone or something else," even though they are completely alone in a hermetically-sealed, soundproof room. For most, the experience is no more detailed than that, but some come out of the chamber claiming to have been in the company of a well-known religious figure. Typically, a figure from a religion they are familiar with: good Catholics see flickering visions of the Virgin Mary; Muslims have met the Prophet Muhammad; and Canadian Native Americans have been visited by certain "spirits." There is even a report of a man who was convinced that he had come into the presence of the Christian God.

But why poke and prod religion? Why not just leave it be? Religious feelings should be respected, people say, and religion is a matter between an individual and their God. But that old cliché doesn't hold water and never did. Religion is a battleground, and today it is *the* battleground. We hear it in the battle cries of fundamentalist Muslims and the rhetoric of America's politically-powerful Christian right. And of course the religiously-flavored confrontations whether they take place in Europe, the Middle East or the Midwest all have political elements in them. But, underneath there is a basic difference that has nothing to do with East and West, Islam and Christianity, Israelis and Palestinians or liberals and pro-lifers, but with two fundamentally different ways of explaining the world and living one's life.

It is the religious approach as opposed to the scientific. These two approaches *are* essentially different and in mutual opposition. One value system encourages faith, acceptance without question, obedience to traditional dogmas; while the other is based on chronic curiosity and the need to verify, to produce evidence for the claims you make. Two so apparently opposing value systems cannot live in peace, side by side. A collision is inevitable, such as when religious dogmas are contradicted by scientific discoveries, or when the latitude for scientific research is constrained by attitudes based on interpretations of Holy Scripture.

In other words, we are dealing with colliding worldviews. Intelligent Design is the label for a doctrine that has its source in a Christian vision that the world was created by God. However, its proponents present ID as a scientific alternative to the theory of evolution and Darwin's theories of natural selection. The theory of evolution lacks explanatory power, they claim, because there are phenomena and observations that cannot be explained through random mutations and natural selection. They are too complex.

The movement's favorite example is the bacterial flagellum, a tail that uses a nano-sized protein motor nestling at its base to move it and propel the microbe around. It is claimed that this mechanical wonder has exactly the parts it needs to function, no more, no less – and cannot possibly have arisen by itself through blind evolutionary processes – so there must have been an intelligent designer behind it.

The designer with which the ID proponents operate apparently got the ball rolling at some point in the distant past, when he invented some extremely clever fundamental structures of life, including these fantastic flagella. And only thereafter was creation left to what we call evolution – to a certain degree.

They are not so enthusiastic about putting a name on this grand intelligence. They would prefer not to speak directly about the Christian God but stick to a discussion of some higher executive power. The truth, however, is something else. As Barbara Forrest, a philosopher at Southeastern Louisiana University, documents in her book *Creationism's Trojan Horse*, the ID movement grew directly out of good, old-fashioned Creationism; people with Bible in hand who assert that the Earth was created 10,000 years ago by a God who took six working days to put together all living creatures in their present form.

The Central Command for disseminating the idea of intelligent design is the conservative think tank Discovery Institute in Seattle. The institute is led by evangelical Christians and its considerable economic support comes primarily from patrons and foundations belonging to the same camp. Its affiliations also appear very clearly in the now notorious *Wedge Document*, a relatively brief, secret manifesto that was leaked in 1999 and placed on the Internet.

"The proposition that human beings are created in the image of God is one of the bedrock principles on which Western civilization is built," they write in the opening sentence of the manifesto. In the second part, they lament the prevalence of scientific materialism and declare war against it. "[T]hinkers such as Charles Darwin, Karl Marx, and Sigmund Freud portrayed humans not as moral and spiritual beings, but as animals or machines ... whose behavior and very thoughts were dictated by the unbending forces of biology, chemistry, and environment .... The cultural consequences of this triumph of materialism were devastating .... Discovery Institute's Center for the Renewal of Science and Culture seeks nothing less than the overthrow of materialism and its cultural legacies."

A "wedge" must be driven into the tree trunk of rationality, the document further states. And an attempt would be made to force intelligent design onto the curriculum of the American education system. Public schools are the preferred battleground – and, in the State of Kansas, it is well known, the combatants regularly wind up in court.

We have grown used to linking this conflict with the US but even secular countries like the UK, Denmark, and Holland have had proponents of the movement suggesting that ID be taught in schools.

In response to the attacks, science has begun to fight back. Researchers defend their fields and their methods in public debates and, as something entirely new, they have begun to write books that look like calls to arms. In 2004, a young American philosopher, Sam Harris, published *The End of Faith*, which wound up as an astonishing bestseller. Originally, he had difficulty finding a publisher, because it was unprecedented in its aggressive critique of religion as a phenomenon. All religion is in direct conflict with reason, says Harris. He deals blows to a range of faiths but in most of the book he rips Christianity apart and bemoans a United States paralyzed by powerful fundamentalist currents.

The year after Harris' bombshell, another American philosopher, Daniel Dennett, published his *Breaking the Spell.* It is a somewhat more temperate contribution to the debate and, according to Dennett himself, it is an attempt to reach religious believers and raise questions about their faith. It does this by examining religion as a natural phenomenon, opening with a very telling illustration of an ant in a field of grass. The little insect toils his way to the top of a grass blade only to fall to the ground again and begin all over. Again and again, it does this. A passer-by observing this behavior would say to himself that there must be some point to it, that the ant is trying to accomplish something or other.

But that would be wrong. You are just seeing the result of a parasite that has burrowed itself in the ant's brain and taken control. The worm-like *Dicrocelium dendriticum* has its own purpose – it wants to find its way into the stomach of a cow or a sheep in order to proceed along its life cycle. To that end, it has taken over an ant and is making it do something that will increase the chances of being eaten by grass-eating cud-chewers. Definitely not in the interest of the ant but quite useful for the parasite.

If you think about it, suggests Dennett, don't religious ideas function in the same way? The idea of eternal life – perhaps even accompanied by harps or willing virgins – can make its carrier blow himself up. Certainly not to the advantage of the martyr but perhaps for the benefit of the idea hitching a ride in his brain?

The British evolutionary theorist Richard Dawkins of Oxford University undertakes the same sort of exposé of religion as a natural phenomenon in *The God Delusion.* The book often takes a biting, sarcastic tone. At one point, Dawkins describes

Yahweh, the God of the Old Testament as "arguably the most unpleasant character in all of fiction" and he argues that religion is a delusion that is not considered a sign of insanity simply because it is shared by a sufficient number of people.

Dawkins has been called the world's most famous out-of-the-closet atheist, and he has long "proselytized" openly for the end of religious faith. In a slightly more organized effort some years ago, Daniel Dennett tried to become the catalyst for a true atheist movement, when he introduced the concept of *Bright*. Dennett pointed out that the only factor that definitively blocked someone from being elected to public office in the United States was godlessness. If you are a politician in God's own land, you have to use rhetoric that includes religion, and you have to have standing in some church. For who could have confidence in a person without faith? In recent years, various opinion polls have made it clear that Americans consider atheism the worst cultural "handicap". People claim that they would rather vote for a gay man or a black man or even a woman before they would vote for an atheist for president.

Even though, today, there are 20,000 members of the Bright movement, it must be said that they have not had much impact and are not much heard in the mainstream. Nor can American Atheists, which fights with legislators and pursues lawsuits for the rights of non-believers, be said to have much more power with its 2500 members and a budget of a million dollars a year. But it looks as though we are seeing the germ of a more aggressive and powerful variant called the New Atheism – with people like Dennett, Dawkins and Harris as standard bearers – which directly and vociferously confronts religion. "It is time to stop pussyfooting around. Time to get angry," says Dawkins in one of his essays.

And for Harris, one of the most important points is to come to terms with the taboo of criticizing religion.

Meanwhile, science has begun putting the phenomenon of religion itself under the microscope. This after science has long kept a safe distance from the religious. Science has, one could say, rendered unto Caesar what is Caesar's and unto God what is God's. Religion and science were considered two essentially different domains that did not overlap, and neither had jurisdiction to make pronouncements about the other. There has been a sort of ceasefire in which people on both sides, politely shying away from conflict, turned their gaze away and more or less tried to pretend the other side did not exist.

But now the shit has hit the fan, and theoreticians and lab scientists are taking up positions on the field of battle. One school consists of Darwinists who claim the existence of religion is based directly on the mechanisms of evolution. In their view, the reason religious conceptions and religious practices exist among human beings and have spread throughout all known cultures is that they provided an evolutionary advantage for our early development. It is an idea that has been elaborated in David Sloan Wilson's *Darwin's Cathedral.* As a biologist, he places weight on the fact that members of the genus *Homo* lived for millions of years, almost up to our own time, in small groups, and religion provided cohesion for these groups.

If individuals are on the same page with respect to a particular view of the world, they are able to feel a kinship with one another and a common opposition to "the others" – that is, groups that have a different view of how the world works. The theory suggests that this clear identification might provide an advantage in battles with hostile groups with a less close-knit community. And it further claims that groups without the capacity for religion would not have a corresponding cohesiveness.

Not everybody buys this idea. There are those who believe it more likely that religion sprung up as a pure side effect of other achievements of evolution. One of them is Dawkins, whose hypothesis is that what in its time provided an evolutionary advantage is a brain with the built-in capacity for creating something so far-reaching and complex as religious conceptions. In other words, religion arose as an evolutionary "by-product" of a well-developed human intelligence. Actually, good old Darwin was on the same track. In *The Descent of Man,* he very briefly examined religion and described it as an excrescence of consciousness itself. Of blood rituals and other superstitions, he wrote that "[t]hese miserable and indirect consequences of our highest faculties may be compared with the incidental and occasional mistakes of the instincts of the lower animals."[1]

One of those who have attempted to formulate a coherent explanation of how religion arises as a "by-product" is the anthropologist and psychologist Pascal Boyer of Washington University. Boyer is one of the stars in a new field known as cognitive religious studies, and his great contribution is a theory that considers religion as an opportunistic mental infection. He lays out his theory in *Religion Explained*, in which he builds on a broad panoply of observations from religious studies, anthropology and psychology. Taken together, they indicate that the way our brains and psychology function makes it terribly easy for religious conceptions to move in. And it is not just about religion in the narrow sense. For the cognitive researchers, Jesus and Santa Claus have the same source, and all ideas about the supernatural are viruses that, so to speak, hack their way into the mental systems we use to understand the world.

You see this, for example, in our apparently inherent tendency to assume that things around us have a purpose. We walk around with a sort of mental purpose detector, which has the effect that we automatically choose to put the world in an explanatory framework that ascribes a purpose to events and figures. The American developmental psychologist Deborah Kelemen has dubbed this "promiscuous teleology." She has observed it in both small and somewhat older children. For example, Kelemen asked four and five-year-old children what various objects – living and non-living – "were for." They accepted the question and answered readily. Thus, a lion is "made to be seen in a zoo," and a cloud is "made to give rain." And if children find a tree suitable for climbing, they don't think it's just an accident that it's in the yard. No, it was *created* by someone and was put there to be climbed in. Young children actively prefer this sort of explanation to "grown-up" explanations about chance occurrences and physical circumstances and, despite influences from the adult world, the tendency continues, according to Kelemen, until the age of nine or ten.

Apart from their promiscuous teleology, children also have an inherent tendency to construct notions of invisible friends who follow them around, says Kelemen. She concludes that from the age of five, they are intuitive theists.[2] They are disposed to ascribe supernatural causes to natural phenomena. It turns out that even the idea of gods and magical creatures comes quite naturally to a two to four-year-old. And anthropologist Jesse Bering of Queens University in Belfast has shown that small children intuitively feel that other people know when they have done something forbidden, even if they did it in a closed room without witnesses.

Yes, but they are only children, you might say. They get over it, once they have acquired reason and maturity. And they do to a certain degree, but the point of cognitive religious research is that, even in our most rational incarnation, we are still hopelessly characterized by psychological tendencies that stem to a great degree from what you could call our social antennae. As a social species, we are adapted to look for causes in the form of social interactions. At the same time, our brains have a fantastically well-developed ability to understand other individuals, to see the world from their point of view and attribute motives and intentions to them. This ability can very quickly create beliefs that there are also motives, intentions and individuals – invisible agents with causal force – behind all the effects for which we cannot immediately see the cause. Religious conceptions fit, hand in glove, with the way we think. Or as Boyer so beautifully phrased it: "Religion is a parasite on our cognitive apparatus."

Now brain researchers are eagerly probing and prodding this cognitive apparatus. With their electrodes and scanners, they have begun to measure the brain in an attempt to understand faith in general, and mystical, supernatural and religious experiences in particular. To understand them in relation to what actually goes on in people's heads. People are speaking of neurotheology as a field of research. And neurotheologists talk about the fact that there are areas in the brain or a network of nerve cells that produce various types of religious experiences.

The most spectacular must be said to be Michael Persinger's attempt to produce mystical experiences by stimulating the temporal lobes. His experiments began in the mid-1980s, when he designed his sound-proof room C002B and constructed his now famous God Helmet. Over the years, he has published a series of articles together with a varying array of collaborators – most of them describing classic double-blind studies in which ordinary research subjects put on the helmet and had no idea what the experiment was about and in which only half of them were actually subjected to magnetic fields.

Persinger works from a sort of unified theory that lumps together a series of apparently different sorts of mystical or supernatural experiences and claims they have the same source. Whether we're talking about visions of hands writing on the wall, encounters with angels or alien abductions, basically it is all about electrical activity in the same spots in the brain. Persinger suggests there is a prototypical experience of what he calls a "sensed presence," that is manifested differently in different people, because they interpret it from a different context. It is a context that turns on culture, psychology and personal experience.

As a person, Michael Persinger is definitely not uncontroversial, and his research doesn't appeal to everyone. He investigates the effect of weak magnetic fields on tissue and cells – particularly brain cells – and it must be said it's a field which exists on the margins of scientific research. We know very little about what effect magnetic forces have on individual cells and even less about what they do to the brain as a whole. Of course, a growing interest can be detected in using powerful magnetic fields – so-called TMS or transcranial magnetic stimulation – for the treatment of, among other things, depression. But magnetism as a phenomenon is not at the moment something that interests cutting-edge research groups or attracts huge grants.

In 2004, Persinger got into a commotion – in print, of course – with a Swedish group led by Pehr Granqvist, which is the only group that has attempted to repeat his results with the God Helmet. Granqvist claims that what determines whether the research subject senses a presence is exclusively dependent on how psychologically impressionable he or she is. In other words, it isn't the magnetic fields that are doing it. Nonsense, Persinger pushed back, pointing out that the Swedes did not use his equipment correctly and did not create the same patterns in the magnetic fields that he himself produces.

We also know that the Sudbury professor is not alone in being able to conjure up a presence. In the fall of 2006, a Swiss research team led by Olaf Blanke published a sensational article in *Nature* in which they described a group of patients with severe epilepsy who were to have the diseased tissue surgically removed. Before the operation, they had electrodes implanted in order to demarcate how much tissue the surgeons could remove without compromising vital functions. But when the electrodes were turned on, something unexpected sometimes occurred.

One woman, whose electrodes were placed on the left side of the cerebral cortex, reacted by looking over her right shoulder every time the doctors stimulated the electrodes. She sensed a distinct dark shadowy figure under her bed and she had a feeling that it was up to no good. Olaf Blanke remarked that it is tempting to invoke something supernatural, when this sort of thing occurs, but it is really only a demon from the depths of the brain.

Whereas Persinger actively provokes experiences that can be interpreted as religious phenomena, others have been more interested in what happens when people themselves cultivate religion and experience spirituality. One of the seasoned researchers is

Andrew Newberg, who is the head of the Center for Spirituality and the Mind at the University of Pennsylvania. Among other things, he has conducted an extensive series of studies of people from different faiths, as they are engaged in various spiritual activities, such as speaking in tongues, or glossolalia as the psychiatric literature has it. This is a trance-like state in which the speaker babbles a stream of apparently meaningless sounds and is ostensibly in direct contact with the divine. Speaking in tongues has been known and cultivated in various religions over thousands of years. Today, it exists in certain Christian sects such as the Pentecostal movement, in which it is a common aspect of worship for members of the congregation, helped along by song and common prayer, to erupt in a stream of incomprehensible babble.

It was in one of Philadelphia's Pentecostal congregations that Newberg and his colleagues found the research subjects for their experiment. Five women volunteered to sing psalms and speak in tongues while undergoing a SPECT (single photon emission computed tomography) scan to measure their brain activity. First, the women were tested to ensure they were free of psychiatric illness and, when that was out of the way, they got down to singing gospel for about an hour. When the last note died down, they were immediately scanned for half an hour, as the researchers were using the singing activity as a control for a comparison with speaking in tongues. This came right after the singing. The women were taken out of the scanner and brought to an adjacent room, where they began to sing again but quickly shifted over into speaking in tongues. Once they had done that for five minutes, they were led back to the scanner.

The images of their brains revealed that speaking in tongues – compared to ordinary hymn singing – showed up as activity in the parietal lobes toward the back of the brain, while there was a marked reduction in activity in both frontal lobes.[3] As Newberg afterwards said to *Science*, "The part of the brain that normally makes people feel in control has been essentially shut down."

Interestingly, the change in activity in connection with speaking in tongues was largely the opposite of what happens when people are in a religious meditative state, which Newberg had previously examined. He used SPECT scanning to measure brain activity in meditating Tibetan Buddhist monks and Catholic Franciscan nuns in prayer. He has described how their practices are accompanied by a typical pattern of brain activity that fits well with the experiences they report.

One of the most conspicuous things was the difference between the brain activity in a monk before and during deep meditation. While the scanning images before the session kept to a palette of reds, oranges and yellows – that is, high activity – a distinct blue spot appeared during meditation. There was a fall in the activity in the left posterior superior parietal lobe. Among other things, the parietal lobe keeps us feeling properly oriented in space. It integrates a mass of sense information and forms representations of the environment, so we can maneuver our way around in it. According to Newberg, this fits with the feeling the meditating or praying person has that they are at "one with the universe." When the parietal lobe is not processing sense impressions from its surroundings, it will feel as if there is nothing outside – there will be a sensation of merging into everything else.

Newberg himself has described the SPECT image of the parietal lobe as "a photograph of God," and, from time to time, you see talk in the media about a "God module," the "God spot" and the like. But more recent research indicates that this conception is much too simple. It is not a matter of just tickling a collection of nerve cells as if it were a button you could push. At any rate, Mario Beauregard of the Université de Montréal has shown that mystical-religious experiences can draw on a large network of brain areas. Beauregard, who classifies his research as "spiritual neuroscience," studied a group of Carmelite nuns at the end of 2006. They were put into a functional MRI scanner, where they felt a *Unio Mystica*, a union with God, as they were being measured.[4] The scanner, which registers blood flow as an indirect measure of brain activity, showed that the nuns had increased activity in widely-scattered areas, involved in processing vision, emotion, body sense and self-awareness.

Advanced scanners are the preferred toy these days, but the anchoring of spiritual experiences in the clammy coils of the brain goes further back than the invention of this expensive tool. Among psychiatrists, it has been well-known for decades that patients with what is called temporal lobe epilepsy have a tendency toward intense religiosity. This is not epilepsy as we usually think of it, where patients lose consciousness and collapse in seizures. Rather, the attacks express themselves as an intense focused electrical activity in the brain's temporal lobes, the part of the cerebral cortex that spreads from the temple to just behind the ear.

One of the first professionals to study and describe the phenomenon systematically was the late American neurologist Norman Geschwind, who published a classic series of articles in the 1960s and 1970s. Among other things, he described how patients develop a characteristic personality that is deeply influenced by religiosity. Indeed, extreme religiosity is a character trait common to these people, but it expresses itself differently depending on where they live. In the US, where there is a wide range of religions, patients with temporal lobe epilepsy often convert from one faith to another at a rapid pace. And a number of them become lay ministers.

In addition to this peculiar personality, there are also religious "attacks." Neurologist Vilyanur Ramachandran from the University of California in San Diego has described how a quarter of his patients with temporal lobe epilepsy reported deeply-moving religious experiences. Experiences that always appeared at the same time as epileptic activity could be recorded with an EEG. They might be intense experiences of bliss and ecstatic feelings from being one with the divine, or the experiences might have the character of revelations with visual and auditory hallucinations. In general, the patients' sense of time and space disappears while the activity is underway, and they usually describe their experiences as sublime.

You can get an impression of how sublime by reading the letters of Fyodor Dostoevsky. He was himself an epileptic and, in one description of an attack, he writes to a friend: "I don't know if it lasted a minute, an hour or a month, but I know I would give the rest of my life to experience it again."

Even in his fiction, Dostoevsky, who suffered from classic *grand mal* seizures and what people have later been able to diagnose as temporal lobe attacks, made frequent use of his experiences. A number of epileptic characters who have religious visions appear in his novels. Prince Mishkin in *The Idiot* is a prime example, as is the middle brother Karamazov, Ivan Feodorovitch.

Most religious people would object that there is so much more to religiosity than visions and revelations and that the majority of believers practice their religion without ever having an extraordinary spiritual experience. This is true, of course. At the same time, it must be said that, in spite of this, such extraordinary spiritual experiences are key in religious history – because accounts of *someone's* spiritual experience are central to pretty much all faiths.

One need only look at some of the people who founded or reformed a religion. The Prophet Muhammad had the entire *Qur'an* delivered to him directly from Allah in the form of revelations and used them as the basis of Islam. The apostle Paul was struck down by a vision on the way to Damascus and immediately began to organize Christianity into a true church. In India, Siddhartha Gautama had mystical experiences in meditation leading to his enlightenment and the realization of the Four Noble Truths; as the Buddha he taught these to others, leading to the foundation of Buddhism. In more recent times, Joseph Smith, founder of the Mormon Church, claimed to have had some rather interesting experiences. He is supposed to have met an angel named Moroni on several occasions and built his church around the flood of revelations he apparently received. This direct contact with God was so significant that the Mormon Church is still saddled with the convention that major ecclesiastical decisions can only be made on the basis of relevant revelations that come to the supreme head of the Church.

Prominent neurologists have from time to time attempted a *post mortem* diagnosis of well-known religious personalities, and many of them had experiences that fit well with the diagnosis of temporal lobe epilepsy. Others find God after receiving a blow to the head. For example, Ellen White, who founded the Seventh Day Adventist Church, experienced brain damage as a nine-year-old that drastically changed her personality. At the same time, she began to have major religious visions.

Encounter in a Soundproof Room

"Trying to understand mystical experiences without having one is like a eunuch trying to understand sex," the American writer on science and religion, John Horgan once said. He may be right; I don't know. Myself, I have never even been close to an experience that could be interpreted as mystical, but now I'm on the outskirts of Sudbury ready to give it a chance.

This is the location of Laurentian University's isolated little collection of buildings. The quiet campus is in an idyllic rural setting, the trees surrounding the buildings are showing the bright yellows and nuanced reds of autumn leaves, and there is a view of a still lake. An academic oasis. It's Friday afternoon, but Persinger warned me that, if I wanted to commandeer his time and participate in an experiment, only weekends were available. During the week, the professor only appears at the university outside working hours, in the evening and at night, when his co-workers have gone home.

During the day, he tends to his neuropsychology practice in which he treats patients and also earns the money to pay for his research. Once in the 1980s, he received $10,000 from a private source who was interested in the effect of magnetic fields on the brain, and there was also a single smaller grant from a private

Canadian foundation. Beyond that, the professor himself has financed the work in his laboratory, while the university has made contributions in the form of his salary and the general infrastructure. It sounds almost like the life of one of the gentleman scientists of the eighteenth century, who as a matter of course used their own fortunes to promote science. This, however, is unheard of in modern research.

"Yes, but it *is*!" exclaimed Persinger, when I first called to pester him about getting access to his laboratory. "But this way I'm completely independent. I could easily apply for public grants, but they all put some form of limitation on what you can use them for. What you can and can't study. And in these times, when research is getting more and more political, the latitude given a researcher is becoming narrower and narrower."

Not all the locals are equally enthusiastic about Michael Persinger doing research to show that religion is an interesting electrical artifact. Some people find that sort of thing offensive. But Persinger is the sort of man who does not take orders. The reason he – as an American born in Florida – wound up in the cold climes of Canada is that he refused to go to Vietnam as a draftee.

"Professor Persinger is always ready to stand up and defend his views," says Vivien Hoang, who meets me in parking lot four. Young Miss Hoang, whose parents actually came to Sudbury from Vietnam, is a Ph.D. student and a member of Persinger's twelve-person team.

"Sudbury is a little town and very conservative," she explains. "Professor Persinger is without doubt the most famous academic we have and the administration doesn't hesitate to use him to advertise the university when it suits their purpose. But as a rule they make life for him and everyone in the group miserable because of his research on religion."

It turns out the group works in a gloomy cellar. On our way through the labyrinth of passageways, we pass by an open doorway through which we see three of Vivien's fellow students, camouflaged in smocks and masks, conducting autopsies on a bunch of lab rats.

"It's cancer research," explains Vivien. One of the cloaked students waves to us with a scalpel in his hand.

Around the corner hangs one of the university's recruiting posters with the slogan: *What are you thinking?* We take a sharp turn into a room that proves to be the heart of the group's work – the joint office where people hang out, read, write and talk. The only technical equipment is a computer and a coffee machine – the rest is paper. Books, journals, stacks of data, anatomical drawings of brains and, on the wall, a collection of newspaper clippings with articles about Michael Persinger.

The atmosphere is relaxed and, at the same time, boisterous, like a dorm room in which every newcomer is quickly caught up in the free-flowing conversation. Vivien has never tried the helmet but would very much like to. Paul Whissell, a young man with long, dark hair who is almost finished with his Ph.D., participated in a single experiment a few years ago, but he discovered afterwards that he was part of the control group that had no magnetic pulses sent through the brain. Stories fly across the round table about experiences they have heard described.

"I heard someone say that she sensed her cat was in there."

"Some people have terrifying experiences, while others say it's wonderful. White light and that sort of thing."

"Once there was a Japanese lady who sat there with tears streaming down her cheeks, jabbering away in Japanese. She said it had been sublime."

"Hello!" Everyone falls silent when Persinger sweeps through the door. He has on a trench coat and a pin-striped suit complete with vest and watch chain. He is a dapper gentleman, around sixty, with wavy gray hair and wire-rimmed glasses. He looks far more cheerful and approachable than the few pictures you can find on the Internet suggest he will be.

"Aha, our visitor from Denmark? Welcome. Unfortunately, I'm busy with meetings for the rest of the day, but feel free to keep talking with my students."

"But what about the helmet?"

"We have an appointment for tomorrow. We'll put you in the chamber at six o'clock, but you need to be tested first. So, be here at three."

Then, he's gone.

The next day, at the stroke of three in the afternoon, the professor walks into the office again and asks me when I got up that morning. Around 7:30, I reply and he looks at the wall clock and mumbles, "Excellent, excellent."

"That will fit perfectly with putting you in the chamber at six."

But, first, I have to go through a little psychological testing, and Persinger leaves me with some questionnaires to fill out. There are almost four hundred questions to which I must answer yes, no, or don't know.

The first thing the test wants to know is whether I like to read mechanics magazines. No, that I decidedly do not. Further down, the questions begin to deal with the extent to which I think people talk about me behind my back and whether I always speak my mind in assemblies or remain quiet. In another questionnaire, I am asked whether I believe in the Second Coming of Christ, whether I go to church, and whether I often get the sensation of being outside myself.

Aha, a personality test and a test for religious and spiritual inclinations. I'm ready to go.

"Do you ever hear voices?" I check off "no" but then think that it wasn't for lack of trying. As a small child of four or five, I wanted so much to hear voices. At that time, my mother was a nurse at a psychiatric hospital, and dinner table conversation was spiced with stories of what the patients in the ward came up with. It was always very inspiring and made me lock myself in the bathroom again and again in the hope that mystical inner voices could be heard in the silence. But I never heard anything.

"Have you ever participated in unusual sexual activities?" No! Let's move on.

"Is your father always right?" Here, I'm a bit puzzled. This must have something to do with faith in authority. In spite of every good intention, I have to answer no.

"Finished? Then I'll just take your papers."

Young Linda St. Pierre is a psychologist, an instructor and Persinger's usual assistant on projects involving the helmet. She takes me through the cellar's hallways to the laboratory where everything is supposed to take place. First, we edge our way through a little anteroom, which for some reason is filled with paper. Articles, books, periodicals, some of them open and strewn about. Towards the back through a small connecting hall is a high-ceilinged room that makes me think of an eccentric second-hand shop. At any rate, I get the strange feeling of stepping back a few decades in time. The interior is a peculiar blend of the worst trends from the 1950s, 1960s, and 1970s. The wall-to-wall carpeting is an indeterminate brown, and there are three small, identical metal desks side by side. There is a lamp on the middle one that spreads a little cheer.

"I brought that," says St. Pierre, who couldn't stand the seventies monstrosities that had been standing there for years.

"Very cozy," I reply. There are bookshelves up to the ceiling all the way around, stacked with dark brown box files and, on the lower shelves, a throng of objects fight for place. I study – almost devotionally – an EEG apparatus that looks like an antique. I'm told that it is forty years old but works impeccably – in fact, it is even more reliable that the modern ones.

On the wall beside the door – which is itself a light institutional green – is a surprising touch of decoration: a drawing of a vase with flowers, done in pastels. The picture is crooked, and I can't make out the signature. The only other decoration is the half dozen diplomas with the name Michael Persinger printed in a flourishing typeface. There is a Ph.D. degree, a license to practice clinical psychology, and several others. There is a collection of scents in a wooden box – probably to be used for sense experiments. I am walking quietly around, when Linda suddenly shouts from an adjoining room.

"Don't look in the chamber. I promised Persinger that you wouldn't see it before you are shut in."

I'm supposed to sit in there with the helmet on for an hour and a half. It sounds like a terribly long time. But first she wants to test me some more. I have to repeat a series of numbers she enunciates with a slow diction, and it goes fine as long as there are only three or four. It doesn't go so well when there are five or six, and I feel like a school child who can't pass an entrance exam.

Then there are some other tests – following her finger over a row of blocks. Again, five or six are apparently too much for my brain. "I'm afraid I'm over forty," I say apologetically. Linda just smiles professionally. Then she gives me a set of earphones and hooks me up to a tape recorder. There is a woman's voice speaking words into my ears, which I am to repeat to Linda and say in which ear I hear them. I do splendidly, I think, and regain some self-confidence. Then the voice suddenly speaks two different words, one in each ear at the same time, and Linda wants them both. Concentration is required.

"Lace and correct," I repeat, while she notes it on her notepad without expression. There is a break and, over Linda's shoulder, I catch sight of a brain floating in a glass tank filled with a yellowish liquid. With my experiences at Harvard's brain bank fresh in my mind, it's like running into an old companion. I notice that I'm smiling at it.

"You need a blindfold on for the next one." I'm blindfolded with an old eye mask – the sort you get on transatlantic flights, but it is stuffed with Kleenex, so not a single ray of light can get in.

In front of me is a board with different-shaped holes cut into it and some correspondingly shaped blocks, which I am supposed to fit into the holes with my right hand. Linda starts a stopwatch, and I slog away with the blocks, which don't want to go into the holes. It's enough to drive me crazy, like being reduced to the state of an awkward kindergartner. I get frustrated, but finally set them in place and am about to take the blindfold off.

"We'll take it one more time with the left hand," says Linda in a flat tone. We do, and the only difference is that it's even more impossible this time and my frustration grows even greater.

"That was good. This last time, you can use both hands."

Later, Persinger breezes through the door with long, swift strides, meticulous in his appearance. The man seems indefatigable.

"How are you?" he booms.

"I'm exhausted."

"Good, good! Exhausted is good. We can use that," he says, sitting down at the table to leaf through my test results. Illuminated from below by the small desk lamp, there is something diabolical about the well-dressed professor.

"Hah. Fantastic," he says after a few minutes, rubbing his hands together. He skims a graph and smiles widely.

"This is wonderful. Do you realize you are in the highest percentile for women with respect to self-assertiveness?"

"Really? I had no idea. ..."

I neglect to mention that my partner has frequently said something along those lines but not exactly with the same enthusiasm.

"And you're also imaginative and creative, unconventional, and, I see, social and extroverted. The perfect combination."

For what?, I think, hoping that he won't look closer at the one with the numbers that had to be repeated backwards. Or, at least, won't mention it.

"But what really interests me here is the lateralization, whether there is a big difference in how well the two halves of your brain function together. It all looks pretty good. Shall we go in?"

Finally. We go into the chamber, the famous room C002B, sound-proof with heavy double doors, box-like and low. There is a thick, yellow shag pile carpet and a velour easy chair with matching foot rest, both partially covered by a white throw-rug.

"Sit down and make yourself comfortable. Comfort is crucial."

Linda St. Pierre is running around outside. Then she comes in to attach the electrodes with something sticky that looks like chewing gum. The two researchers speak over my head, some technical details I don't grasp. They are having a hard time making the electrodes stick, because my hair is too thick and too unruly. More chewing gum. Finally, it works. They jointly lift the helmet and, as if it were a crown, set it gently on my head. Persinger looks at me critically from the front and from the side.

"It seems okay. Yes, it fits so you get the fields just over the temporal lobes."

The professor positions himself in the corner and takes a couple of pictures with my camera. Then, some swimming goggles, lined with Kleenex, are put on me and taped to the helmet.

"Is it completely dark now?"

I give my word that I can't see anything, but then Linda shouts from the control room that Persinger had better come and look at my EEG. They murmur together for a moment. Then the professor comes back in.

"You need to relax."

"But I'm completely relaxed."

"Your brain activity shows something else. Stop thinking so much."

They shut the first door, then the second door, with a smack. A bit later, there is a scratchy sound from the speaker mounted in the corner.

"Dr. Frank, are you ready?"

"I think so ..."

"Then, here we go."

There is a hollow sound, when they shut off the microphone, and I'm alone. Initially, it feels like submerging in a tank of water – only the tank is filled with an impenetrable darkness, rather than water. The only thing I can hear is my own breathing. I try to relax and think about how the whole scene must look. A big, yellow crash helmet with a couple of blue-glassed goggles glued on and a thick fleece jacket to keep warm. An absurd experiment. So, I think a little about everything I've heard about the experiments and can't quite fathom that nothing really happens – yes, a few lights flicker and, yes, there are some small sounds like a cricket down near the left corner of the footstool, but nothing that would count as a religious experience. So, the time goes by in the dark, and my thoughts stream along, even though they are a little sluggish. *Klonk*. The loudspeaker startles me.

"That's the end of the first phase; I'm coming in."

Persinger squats down – he's remarkably agile for a sixty-year old – and he hands me a questionnaire. I answer no to pretty much everything. No, I wasn't dizzy; no, there was no strange smell, except for the stuffy room and the shaggy carpet, but presumably that doesn't count. Persinger takes the form with him and closes the double doors again.

I stuff the tissues back under the goggles and feel a certain disappointment. But okay, at least, I'm in good company. Linda St. Pierre told me that Richard Dawkins himself, the world-famous evolutionary biologist, was here a couple of years ago with a film crew from the BBC and he didn't experience anything either. Maybe, you have to be a little religious already to be able to imagine something.

"All set in there? Here comes the next pulse. And remember, Dr. Frank, these are subtle effects; it's not *virtual reality,* this stuff.

Try not to categorize what you're experiencing. Just sense what's happening."

It quickly becomes clear that this second pulse is doing something. The whole room seems different. At first, it's just a sense that I am more acutely aware of things and the atmosphere feels like it is charged. Then there are sounds. Some clear sounds that can't be coming from me. It's not my breath, but something else entirely, sort of unsettling. It's as if something is creeping or, rather, crawling around in front of the footstool.

It's like there is a body behind the sounds. Yes, there *is* some sort of creature in the room, and it is not a particularly pleasant creature to have as company. It is possible that other people have been in contact with prophets and blessed virgins in here, but not today. He – yes, it's a he, if I have to ascribe a gender – is something eerie and threatening. A little like Gollum from *The Lord of the Rings*. My eyes shift, as if they are trying to follow the sounds, and I have a feeling that there is something over to the left. At the same time, there is a classic fear response that grows up from my belly and envelops my entire body. One of those where stress hormones are pumped into the blood stream. I want to tell them to stop and ask them to come in. It feels like he/it might reach up and touch me at any moment – I want badly to pull in my hands and feet but remain where I am. An image appears in the back of my mind – the needles attached to the EEG machine on the other side of the wall must be going crazy, swinging wildly from side to side, tracking a curve of hysteria on the scroll of paper.

Easy now, it's just an experiment. There's nothing out there in the dark, not really.

Slowly, these foggy impressions fade, the thing on the floor stops moving and is finally gone. I try to control my breathing and concentrate on relaxing. But it's not long before things start to go wrong again.

"My hand ... it feels weird." My voice is thin and weak, with no resonance in the soundproof room. But I can't shut up, because my right hand has been screwed off and is at a right angle to my arm. I know logically that it is still on the arm rest, pointing away from my body, but it feels as if it is turning toward my stomach. At the same time, there is a force pressing on the palm of my hand – stronger and stronger, actually, as if somebody were standing there pressing it. Then, things start happening with the left arm and, finally, they are both bent at the same angle.

I don't move. I just let myself feel, as the professor has asked me to do. It is a disgusting feeling. A bit later, both arms go up until they are hovering up around my ears, as if the arm rest were up there as well.

Finally, Persinger comes in with his questionnaire. The lights come on, and I can just glimpse some light, which must be the professor, who once again takes a picture. The awareness that I'm sitting in an armchair wearing a crash helmet with goggles covering my eyes becomes quite pressing. So I help take off the helmet. Cautiously, I peel off the electrodes and leave them in the chair. I leave the white paste, which has now stiffened into plaster-like clumps, in my hair for the time being. The last questionnaire has to do with my final assessment of the experience: quite unpleasant.

"Sit down." Persinger does not look at all dissatisfied, and he opens the table up for questions behind the glow of the lamp. He leans forward over one of the side-by-side desks.

"You say it was eerie? Imagine how these sorts of experiences would seem if they took place spontaneously at three o'clock in the morning, while you were lying in bed. And how you would feel if your friends told you it was probably Satan coming to get you."

Yes, I can imagine it, no problem. It's not like you can just shake off the experience. Even here, afterwards, my voice sounds thin and uncertain.

"It felt exactly like there was something crawling back and forth."

"Yes, it's not often people talk about movement, but it was probably due to the fact that your brain was so active, which we could see on the EEG."

I assure him that it was quite impossible just to lean back and observe it – "it was scary," I say, sounding like a child.

"Yes, I understand. But at the same time you were trying to localize what you were feeling, you changed the activity pattern of your brain and the way the activity in the cells interacts with the magnetic field. It's not that unusual. Around every tenth subject reports that 'it' moved to the side, when they tried to focus on it. And you can imagine how the movement gives people the feeling that there is a living creature out there, actively trying to stay outside their field of vision."

A creature, maybe, but a direct religious experience it wasn't. No blond, Scandinavian Jesus in radiant light and white robes, no holy prophets, not even something that could be interpreted as a spirit.

"Eight out of ten people who put on the helmet have an experience of an altered state of consciousness, but there are a number of parameters that determine *how* it is manifested. In the first instance, it has to do with how sensitive the temporal lobes are in the individual. It's not the same for everyone. And then there is a huge element in the form of the cultural baggage you come with.

Both in relation to the expectations you have, how you categorize unusual experiences in general, and in particular what religious tradition you come from."

Atheists typically reason that what is happening is a little trick their brain is playing on them, and they can distance themselves from the experience. Then there are research subjects with a Native American background who refer to a spirit or long-dead ancestor, and others who attribute the presence to their own particular faith system.

But what is it that's going on? That's not yet clear to those who actually study the functions of the brain. Michael Persinger is working on the basis of a hypothesis that we don't just have a single "self" or sense of self but several and that they, so to speak, keep to or are created in different places in the brain. You could also talk about different aspects of the self and that aspects we do not normally view as independent suddenly seem to function that way in certain circumstances.

"I look at it this way: Our normal sense of the self – what we usually describe as 'me' – is connected to the left hemisphere of the brain. This is where a lot of linguistic activity takes place and the sense of the self is a very linguistic phenomenon. The right hemisphere, however, has its own counterpart to the left's sense of the self, but in an alert state, it is inhibited or repressed by the communication that goes on between the two hemispheres. When for one reason or another the proper conditions are created, this right counterpart can intervene in consciousness and seem like an 'other.' This is what we call the sensed presence. The 'creature' you experience in the helmet."

This is the theory, anyway. Or one theory. It fits with the hypothesis of the well-known neuroscientist Michael Gazzaniga that the left hemisphere is the haunt of what he calls the "left hemisphere interpreter;" a network that tries to keep us together as an individual by constantly interpreting the information that comes in from all the brain's other networks and makes it into a coherent personal story. The narrative of the self, you could call it.

But everything about this vague thing, the self, what it consists of, where it resides, and how it is constructed is still entirely speculative. And as for the pulsating magnetic field, no one knows how it works on the cerebral cortex. Years of experiments on people and animals have shown that the magnetic field used is significant. The fields Persinger uses are no stronger than those generated by an ordinary computer screen, and few people have experienced God or the Devil in front of their computer. Over the years, the group at Laurentian have experimented with innumerable variations in magnetic fields, and it looks as if there are specific patterns of pulses that work and that their intensity is decisive. What happens in the helmet is that, for the first twenty minutes, you get a weak pulsating magnetic field over the right hemisphere of the brain and then twenty minutes when the field is directed toward the temporal lobes on both sides. This special pulse was developed by a former student and is called the Thomas pulse.

But that there are other effective pulse patterns can be gleaned from an article in the journal *Perception and Motor Skills,*[5] in which a strategically placed clock radio evoked visions of the sacred. The article describes the case of a young Canadian woman who complained about inexplicable nightly visitations. At one point, the woman began to be visited by an invisible presence she identified as the Holy Spirit himself. He did not come to proselytize or to speak to her but apparently to repeat his earlier success in instigating pregnancies and virgin births.

The woman could describe his persistent visits very exactly. They started typically with a feeling that her bed was vibrating violently. Then, she sensed a creature moving slowly down her left side and through her vagina into her body, where it usually took up residence in her uterus. After some moving about there, the woman then felt the presence of an invisible child who seemed to float over her left shoulder.

She eventually grew exhausted and tired of these repeated night visitations and sought medical help. The doctor called in Persinger, who measured and analyzed the magnetic fields in the woman's bedroom and found that her clock radio, which was very close to the headboard of the bed, emitted a magnetic pulse. This pulse corresponded exactly to the so-called 4 microTesla pulse, which can be used to provoke epileptic attacks in rats and especially receptive people. When the radio was removed, the visitations stopped.

However, Persinger twists in his chair a bit when I raise the question of a more concrete mechanism.

"The effect presumably makes some nerve cells become active and emit electric signals to which other nerve cells react.

We know that our weak pulses and fields produce currents that translate into a microvolt and nanoamp magnitude in the cerebral cortex. This is the effective area for many current effects on tissue. We also know from experiments with rats that certain effects can be blocked by chemical substances that bind with special receptors on the surface of nerve cells. The same receptors that react to morphine and other opiates."

On the other hand, Persinger and his colleagues believe that the general effects experienced as a research subject depend on the fact that large groups of cells take part in new activities.

"We put vision and hearing out of service in the darkness. So the millions of nerve cells that are normally occupied with servicing these two primary senses don't really have anything to do. This means that they can be recruited for subtle magnetic field activities by our equipment."

Persinger has an idea that it must be the same effect at work for shamans and otherwise spiritually inclined people of all cultures who go to remote caves and similar isolated places to achieve contact with the divine.

"Muhammad actually went into caves to receive the divine word. There is the wandering in the desert tale of Christ, where he fasts for forty days and meets an incarnation of Satan. All sorts of cultures have cultivated the practice of going to an isolated place and indulging in sense deprivation."

An ironic smile plays around Persinger's lips. "We used to call our room in there Muhammad's cave, but that was before September 2001..."

Belief vs. Knowledge

"Of course, the most important thing has to be to determine how religious phenomena have arisen. There must be a reason why people in all ages and all cultures have formed conceptions with a clear religious content."

Thus wrote the philosopher and theologian Johannes Sløk in his book *The Religious Instinct*, and today it can be said that the biologist has a tentative answer. As a good atheist, I can't help speculating about what the new knowledge we have derived from brain research in particular will mean for religion. Will it be driven from the market or will faith be able to hold the fort?

The existence of God does not become more evident with time. To the contrary, the data is overwhelming, indicating that the sacred is found between the ears. Evidence comes not only from the yellow crash helmet, but from the psychiatrist's clinical descriptions of hyper-religious epileptics with overactive temporal lobes. Moreover, current research is increasing our understanding of how our own brains create a broad range of spiritual mental states and our understanding of how our own brains create what people have traditionally believed came from the outside. Electrifying brushes from God's finger, the wings of angels and the breath of the grand inconceivable universe have gradually come to be seen as electricity in a special pattern of synapses between brain cells.

What will happen when all this deeply fascinating academic knowledge finally seeps out of the ivory tower, and becomes general knowledge? Can detailed knowledge about how the brain works and well-established theories about how religion arose in the evolutionary process change people's attitude toward religion in practice?

Or to put it more bluntly: can knowledge cure faith?

Forget it, some will say. Faith in reason is the most naive faith of all. Knowledge and rationality simply don't have the teeth to penetrate the elephant skin of religion. Just look at history, they will go on, pointing out that since the Enlightenment there have been scientifically-minded individuals who have stated with self-assurance that faith and superstition – whether in heavenly gods or subterranean gnomes – will inevitably fall before the unstoppable march of science and reason. Unfortunately, they have not proved correct. Even today, after three hundred years of almost inconceivable scientific and technological progress, a significant proportion of the population still have a rock-like belief in virgin births, the sulfurous fires of hell, and a paradise reserved for martyrs. And just look at what has happened in recent decades – instead of receding, religion has conquered a good share of the market.

Not even among those who have science as their milieu and researchers studying religion as a biological phenomenon is there much optimism. According to Richard Dawkins, religion is "comparable to the smallpox virus but harder to eradicate," and Pascal Boyer, who describes religion as a cognitive parasite, has said that he does not believe for one moment that even a well-founded biological explanation of the origin and nature of religion will mean anything for people who buy into the conception. This scrounging parasite is so well-adapted to our mental equipment that it will continue to develop unchallenged in the coils of our brains, generation after generation. It looks as if biologist Andrew Newberg agrees with this interpretation. Along with the late psychiatrist Eugene d'Aguili, he has written a book called *Why God Won't Go Away*. And on his homepage, he provides a brief summary of his position:

> The main reason God won't go away is because our brains won't allow God to leave. Our brains are set up in such a way that God and religion become among the most powerful tools for helping the brain do its thing—self-maintenance and self-transcendence. Unless there is a fundamental change in how our brain works, God will be around for a very long time.

But is this true? Religious belief is not unaffected by the progress and circumstances of the world. When you look out over the international landscape, you see that there are great differences in the numbers of believers in various countries and cultures. There has been a number of international studies into this area in the last couple of years, so we know, for example, that more than ninety-five percent of those asked in countries such as Nigeria and Indonesia answer that they have always believed in God. In the West, you only find such a high number in a few countries, which are so-called statistical outliers – extreme cases. In the United States and Ireland, nine out of ten say they are believers, while the percentage in countries such as Denmark, Sweden and the Netherlands, is down to between four and six out of ten.

"There are places where you don't teach religious texts as truth. The Bible is presented as a piece of literature," remarked the philosopher and neuroresearcher Sam Harris, when I met with him to discuss the future of religion. Harris is in his forties, soft-spoken, not given to grand gestures, but behind his calm exterior is a warrior. He says straight out that religion must be fought. "At least, if people are to come out of the twenty-first century in one piece." He stresses that religion is not the only effective mental virus we are facing – we can regard science as another.

"If there is anything that is adapted to our cognitive apparatus, it is science. We human beings have a deep desire for our conceptions of the world to fit with what actually is going on. And history shows that we will move away from religion's explanatory model, when science provides the real causes for events or phenomena."

One of the best examples is the progress of medicine. We are, and presumably have always been, desperate to understand why we get sick and what we can do about it. Before modern science came to the fore, it was religion that offered an explanation. If you fell into a seizure, frothing at the mouth, it was probably because someone had cast the evil eye on you and you had to go to a witch doctor or recruit people to pray for you . If your son or daughter started talking to people nobody could see or hear, they were most likely possessed by a demon, which, of course, had to be exorcised by a priest. The demand for exorcisms and witch doctors, however, fell drastically once the apparently demonic behavior could be explained as epilepsy or schizophrenia or ascribed to some other organic cause. Particularly because these conditions could be treated on the basis of the scientific explanation.

You can imagine a corresponding effect, once science has a really good explanation of spiritual experiences. When we can explain in detail what happens physiologically to a person who, for example, overcomes his egotism and loves his neighbor as himself. As Sam Harris says, it is important for science to try to understand this sort of psychological state, and not simply religious belief but other apparently airy and incorporeal phenomena such as ethics and human happiness.

"For when we do that and when our understanding spreads, the conception of a god or a prophet writing a book about how the world is put together will become pointless," he says. And alongside intellectual understanding, he imagines that neuroscience could contribute a user-friendly technology.

"Imagine if Persinger's helmet really delivered the goods. That not just once in a while but every time with absolute certainty you got a nuanced and powerful spiritual experience that could be delivered in twenty different versions – the Jesus helmet, the Buddha helmet, or whatever is in demand. Then, we would at some point be able to talk exclusively about what goes on in the brain in these states."

Actually, Michael Persinger is of the same opinion. He has seen people disappear into room C002B with a strong faith and come out with serious doubts that have just grown bigger and bigger. And the professor believes that there is a future scenario, in which his technology could be developed into a sort of spiritual aid. A world in which, instead of gathering in cathedrals, temples or mosques, we could go into the bedroom or sit down in a corner of the living room, put our magnetic helmet on over our temporal lobes and have a religious experience. Become one with the universe, meet God, or whatever you want to call it. A practice that will put a damper on our existential angst in a society in which we are becoming ever more individualized and isolated.

One of Persinger's former colleagues is already on the market with a first-generation product. Todd Murphy has created the so-called Shakti, a contraption that can be purchased as a helmet and as a sort of headband that has two, four or eight magnetic poles, which you can move around so they affect different parts of the cerebral cortex. Spiritual technology for altered states, says the advertisement on the homepage for Shakti technology. To date, the product has been sold to several hundred users. Murphy, who today lives in San Francisco, reports that most purchasers have been middle-aged men who are typically out "to investigate consciousness in one way or another."

Shakti is connected to the sound card of your home computer, which controls the magnetic pulses, and the apparatus comes with a manual that recommends where to put the magnetic poles on the scalp and where to be careful. On its homepage, you can see a judicious selection of user testimonials.

"A wonderful experience of well-being and mental alertness," writes one.

"Three days after my last session, I was filled with energy and in a fantastic mood, even though the weather was lousy and everybody else was in a bad mood," reports another. And a third describes how, after an hour of stimulation, he noticed a marked increase in loving feelings then a warmth in his head and chest, upon which he fell asleep.

"Most by far aren't after any exceptional spiritual experience," Murphy explains. "You can easily get them by using Persinger's experimental protocol in which you deprive yourself of sight and hearing impressions. But many are interested in the effects you can have *after* the sessions. They want to feel better."

So, there are special instructions for pepping up your mood by stimulating your left frontal lobe for twenty minutes, once a week for six weeks. This is based on results that Persinger and his former colleague Laura Baker-Price published after treating patients suffering from depression and anxiety after head trauma.[6] The patients reported that their depression lifted concurrently with changes researchers could see in their EEG signals over the six weeks of treatment.

"With time and with more research, I'm sure the technology will be developed in a way that allows it to be directed toward very specific effects and experiences and make it possible for everyone – independent of their neurological constitution – to achieve them. Today, we are limited by the fact that all brains are different and that we don't know how to shape the conditions to fit to the individual's neurology."

When it comes to opportunities for spiritual experiences with magnetic stimulation, Todd Murphy does not believe they will ever replace religion.

"In the sense that people must have an inner paradigm, a framework for interpreting experiences. They have to have a vocabulary with which to express their experiences and the changes they undergo, and religion, which has been here for so long, offers for the time being the best vocabulary. I have seen how even hard-nosed atheists resort to talking about what they experience in religious terms. But okay, when they are to explain the mechanics behind it, they turn to neuroscience. I don't believe science can produce a terminology that is as emotionally stimulating as the religious. We will see a lot of people who intellectually *know* that these transcendental experiences come from their brains but *feel* that they have to do with something else, something higher. There is a domain of knowledge and a domain of emotion. I think people will provide explanations from their knowledge but act from their feelings. And this division is an excellent solution, I think," says Murphy. And you can almost hear the wry smile over the phone as he continues: "As Jesus said, don't let the right hand know what the left is doing."

A strong proponent of spiritual experiences that are chemically cleansed of religious content is British psychologist Susan Blackmore. Blackmore is known for her unusual hair – it has been both green and pink – and for doing serious research into the paranormal. She has had a long university career, but is now a freelance writer and broadcaster. In an interview in *Nature*[7] from 2004, she described how she had had "the experience of a lifetime" in Persinger's yellow helmet; but, even beyond that, she is a veteran when it comes to mystical experiences – "with and without drugs." Susan Blackmore has had out-of-body experiences and experiences of being one with the universe; she has seen the long tunnels and the bright, white light that we typically see described by people who have had a near-death experience, which many of them interpret as the entrance to heaven and irrefutable evidence of life after death. Blackmore disagrees.

"Just to be completely clear," she says. "I am not just an atheist; I believe that belief in God is both wrong and directly harmful. On the other hand, I am convinced that experience with the mystical can be positive, valuable and provide insight."

Blackmore believes it is important for these experiences to be studied quite simply because it may help get rid of religion.

"I would like to see a world in which people have left behind the false and harmful religions we know today, but where they work to attain a mystical insight and understanding. There are mystical states in which you see the world as a whole and experience yourself as a fully integrated part of everything that is. It can be a very positive experience that leads to less selfish behavior and greater personal satisfaction. I wish more people were capable of achieving that sort of experience without thereby encouraging traditional religion."

Religion, as everyone knows, is a mutable phenomenon and, with a bit of good will, you can see some of the traditional interpretations mutate in the direction of something that almost seeks to merge with science. Or absorb elements from it.

This does not apply to the fundamentalists of this world, who stick to their religious texts, whether they are called the *Qur'an*, the Bible, or something else. And when science collides with the approved dogmas of these sacred writings about the world, they simply reject the science. After all, the proponents of intelligent design cheerfully insist that scientists are coming up with pure assertions that cannot ultimately be proven. But outside fundamentalist circles, more modern theologians may be found and, among them, a far more flexible explanatory model may be observed. This is the case throughout most of Europe, where the very definition of the Christian God is becoming more and more indistinct and intangible. People no longer talk about a vengeful father or an energetic miracle-maker who intervenes in world events like a child playing around with an ant farm. Instead, we are dealing with something inexpressible, something that unfolds behind all the reality we can see, hear, and measure, and which embraces everything into its peculiar existence.

People typically say that science deals exclusively with and explains the everyday world, the physical universe. Faith is qualitatively different; it provides explanations on a different plane, which can't be discussed within the scientific framework. But for many, there is room for both.

This sort of dual vision was clear during the 2003 BBC Horizon program on religion and the brain, in which Richard Dawkins and other scientific pioneers were interviewed along with Anglican Bishop Stephen Sykes. The bishop was extremely obliging. He conceded there was a huge difference between the experiences Persinger's helmet could produce and a *genuine* religious experience, and maintained that religion had nothing to fear from neuroscience. As far as the bishop was concerned, it was perfectly fine if science attempted to explain the world and he was happy for it to co-exist with religion.

Apparently, today's theologians have no problem with serving two masters. But the *rapprochement* does not end here. One can actually hear Christians go so far as to argue in favor of their religion because of its utility – it is not about how true the doctrine is but how practical it is. Religion is a good thing to have because it gives life meaning, they say. Some even use this to try to claim that religion is good for your health. People ostensibly have lower blood pressure and a longer and happier life if they believe in something – exactly what doesn't matter, just as long as they *believe.*

The media loves stories about how faith in *whatever* can accomplish something or other that modern medicine can't. And there is no shortage of studies, because it has become increasingly fashionable to study whether members of this or that religious group live longer and healthier lives than people who don't believe or go to church or pray regularly. And indeed, many studies actually find that religious people are healthier than non-believers. But that is not so strange, since funnily enough it turns out that the believers quite simply live healthier lives, generally consuming less alcohol and tobacco, when not renouncing it altogether.

But the studies keep on coming and in the dialogue between theologians interested in health and doctors interested in theology, there is an indefinable mixture of psychology and "spirituality." However, if you delve into the sometimes unctuous rhetoric, there is at bottom the same mentality as in the atheists Blackmore and Harris. It deals with religion as *wellness.*

In this light, you can easily imagine that, in time, "soft" modern religion will be absorbed by the scientific way of thinking. On the other hand, you can just as easily imagine that the fundamentalists of this world will be provoked to re-arm precisely because of this impertinent science.

The signs are there already. You can see religious groups creating unholy alliances, because in spite of the different gods and prophets they all feel more spiritual kinship with each other than with the godless. This family feeling comes out, among other places, in the struggle over the doctrine of evolution. Aggressive creationism and its successor in the movement, intelligent design, have been a Christian specialty until now, conceived and propagandized by Christian fundamentalists in the US. Now they are forging bonds with likeminded Muslim organizations. A prominent example may be found in Turkey, where the *Bilim Arastirma Vakfi* organization is using tactics borrowed from American creationists to fight against Darwin and all his kin east of the Bosporus. And the collaboration goes in both directions. This could be seen in 2005, when the *BAV* flew a spokesperson to Kansas to testify for intelligent design in court proceedings on the extent to which intelligent design should be a part of the school curriculum in the state.

Generally speaking, science keeps its distance from all manner of fundamentalists. But there are forces working for a sort of reconciliation or genuine meeting between the two. One of the foremost proponents of this possibility is Edward O. Wilson, who is professor emeritus at Harvard University and world-famous as the father of sociobiology. Wilson, who has recounted in his memoirs his upbringing in the strict Baptist milieu of Alabama, has since his youth been a non-believer. Nevertheless, he recently joined forces with America's Christian right in a cause. The grand alliance occurred in the struggle for environmental protection – an issue that interests everyone but for which they have different names. As a biologist, Wilson wants to preserve the biodiversity in the world, and the Christians want to protect God's amazing creation.

On the surface, the project may seem harmless. Perhaps, something politically positive may come out of it if the US adopts a greener attitude – even if it is partially on a Christian basis. But as a general model, it cannot be used, primarily because science itself constantly produces new technological possibilities that provoke and incense religious groups. Today stem cell research is an affront to some, tomorrow a new discovery may be anathema to others.

A conflict between the scientific worldview and the fundamentalist belief in God is unavoidable, because both sides claim to represent reality. Having two opposing concepts that demand to be respected but which mutually preclude each other is just not possible.

3

MORALITY COMES FROM WITHIN — THE BRAIN AS ETHICS COUNCIL

"How do human beings decide what is right and wrong?" The question is printed across my computer screen and it is indeed a good one. Where *does* morality come from?

The standard view is that knowledge of good and evil is something you learn from having it explained. Some people stubbornly insist that morality grows directly from religious roots and that it can only be grounded in faith. Others look to philosophy, psychology or sociology, claiming that morality is a set of practical rules handed down by the dominant culture of the time to keep a society from corruption.

But none of these schools of thought really answers how people determine what is right and wrong. We don't really formulate moral rules that we live by, we just seem to have a general sense of what is "right" and "wrong" in a given situation. The knowledge is just "in there" somewhere, like some sort of program that is activated and makes itself known almost more as a bodily sensation than anything that could be explained with words. But what is the mechanism behind this peculiar instinctive morality?

Welcome to the *Moral Sense Test*, which is "A Web-based study into the nature of human moral judgment. How do human beings decide what is right and wrong? To answer this question, we have designed a series of moral dilemmas to probe the psychological mechanisms underlying our moral judgments."

Specifically, they want to know how factors such as religion, culture, age and education affect those judgments, the test designers explain. Then they give me the option of clicking on one of three fields labeled English, Spanish, Chinese. If I am to be quizzed about my moral habitus, it had better be in English. I am then assured that the study is easy, quick and completely confidential and makes me a vital part of an exciting research project.

Who can say no to science? "Just click here."

I click. Before we get down to business I am faced with a number of questions about my personality ranging from nationality and ethnicity to how religious I am and what I do for a career. Gradually, as the series of boxes is filled out, I can look at myself as a series of cue words: first-born, forty-one years old, unmarried, no children, no religion. It makes me a little pensive.

Then there is the last question – which is whether I have ever taken a course on moral philosophy or read any books on the subject. That's a bit embarrassing. I know a little bit about moral philosophy, but if I am to be honest, I've only scratched the surface and could only, if pressed, come up with a quote here or a general principle there. I confess my ignorance, and am finally ready to begin.

The quiz involves some fifty scenarios that deal in one way or another with saving a larger group of people at the cost of a few. I am assured that all the information about my moral judgments will remain between the researchers and me. That's comforting.

> It's wartime, and Jeff and his two children, eight and five, live in an area occupied by the enemy. At enemy headquarters, there is a doctor conducting painful experiments on people, experiments that inevitably lead to death.

What in the world are they up to here? You don't expect this sort of Dr. Mengele scenario from a serious Harvard study.

> The doctor is planning to conduct experiments on one of Jeff's children, but he will allow Jeff to choose which child.

How perverse.

> Jeff has 24 hours to bring one of his children to the doctor's lab. If Jeff refuses, the doctor will take them both and experiment on both.

And then the question.

> Is it 1) forbidden, 2) allowed, or 3) obligatory for Jeff to take one of his children to the lab?

What a scenario! It reeks of twentieth-century horror, and you can't help seeing in your mind's eye an emaciated Meryl Streep in *Sophie's Choice*. But what is the answer? After a bit of agonizing, I guess I'd have to say – allowed. In reality, poor Jeff only has the choice of keeping one child or losing both. I'm not wild about kids myself and have never contemplated having any, but somewhere deep inside me I still feel a certain degree of empathy. I click on the second possibility and move on to the next dilemma.

> Enemy soldiers have taken over Susan's village, and they have orders to kill all the remaining civilians.
>
> Susan and some of her neighbors have sought refuge in the cellar of a large house. Susan hears voices outside coming from some soldiers who have come to search the house for valuables. Susan's infant child begins to wail loudly. Susan covers the child's mouth to stop the noise. If Susan removes her hand from the child's mouth, his screams will arouse the soldiers'

> attention, and they will kill Susan, her child and everyone else hiding in the cellar. To save herself and the others, Susan must strangle her child.
>
> Is the strangling of the child 1) forbidden, 2) allowed, or 3) obligatory.

I don't know. But really, I guess, it's the same as before – allowed. You can't really say it's obligatory for the poor woman to strangle her screaming infant.

The next dilemma, for a change, is not about war.

> Denise is standing on a footbridge over some train tracks and sees an out-of-control trolley headed toward five people, who will be hit and killed. However, Denise can pull a switch and send the trolley down a sidetrack. There is a man on the sidetrack. By pulling the switch, she will save the five on the main track, while the man on the sidetrack will be killed.

This one you can call a no-brainer. Of course, it's morally acceptable for Denise to pull the switch. One person will die, but she will save five. And the trolley theme continues in the next question.

> Frank is standing on a footbridge and looks down to see an out-of-control trolley about to kill five people. Frank knows he can stop the trolley by pushing a heavy object onto the track in front of the trolley. The only heavy object nearby, however, is a man who is standing on the bridge with him. Frank can save the five by pushing this man down from the bridge, which will kill the man.

Okay, this is a little more complicated. Whereas the situation with Denise had something more distanced and almost academic about it, you can feel this one in the pit of your stomach. Imagine standing on a footbridge, watching an approaching catastrophe and realizing your own decisive role in it. You gather your courage, run over toward the heavy-set guy, put your hands against his back, feel the resistance from his weight, and you are just able to see the look of complete amazement, when he turns his head the second before he falls. No, it would be morally indefensible to push a man to his death in cold blood! I click resolutely on: *1) forbidden.*

Morality's modern masters

A few hours after I've finished testing my personal morality and after a trolley and subway trip through Boston to Cambridge, I arrive at Harvard Square. The red buildings in mock Tudor style are not pretty to look at, but what does that matter, when the sun is beaming down through a clear, blue sky? It's an incredibly beautiful day. Steady streams of students glide by on the asphalt path, and it's like visiting a picture-perfect diorama of the American dream. Then, coming directly towards me is a young man in a dark red T-shirt with a mug shot of George Bush on the front. With lips pursed, the President looks like a chimpanzee, and the yellow text beneath the picture reads: *Blame Yale.*

I walk past the large, flat Science Center and on towards the bizarre Memorial Hall, which was erected to honor Harvard's fallen in the American Civil War. William James Hall towers up behind it, a 1970s grayish-brown concrete silo, a thoroughly modern ivory tower. Inside this horribly ugly building, you can find some of today's best known researchers into human nature. There is psychologist and linguist Steven Pinker, who has practically achieved rock star status on the academic scene. There is developmental psychologist Elisabeth Spelke, whose research on small children has led to a breakthrough in the understanding of what kind of basic knowledge we are born with. And there is the man I'm on my way to see, professor of psychology and biological anthropology, Marc Hauser. He studies the evolutionary development of human mental abilities – in particular, he has tried to understand the origin of language through many years of experiments on monkeys. But recently he has devoted himself to morality. Here, his primary tool is the Moral Sense Test, and his research animal is the human being.

The Cognitive Evolution Laboratory is on the seventh floor, and its main office is located snugly in a corner near the elevator. When I enter the office, it is like stepping into an oasis. The corridors of William James Hall are as anonymous and boring as if a bunch of accountants worked here. But Hauser's office is like a living room with warm colors and bric-a-brac everywhere. The surfaces are covered with pictures of animals and an array of African and Asian artifacts. The atmosphere of homey comfort is made complete by an old, long-haired dog – I guess part golden retriever – slinking around with a bandanna around its neck. It looks as if it is suffering from rheumatism and only acknowledges me in passing as it settles with difficulty beneath the round table, where I set up my MiniDisc recorder. I take out the microphone and make sure the recorder is working, while Hauser takes care of some e-mails at his desk.

"Good to see you again," he says distractedly. I had visited him earlier in the year to talk about research into the mental traits of animals. I know from experience that he has a very busy schedule, so it's a question of trying to get everything I need done before I am – politely, of course – thrown out to make way for the next appointment of the day. I also know that Hauser talks so fast that any attempt to follow along on a notepad will be futile. He fires sentences like bullets and you have to watch out you don't get hit. The firing gets underway as soon as he sits down at the table.

"Things are really happening," he says unsolicited.

I can't help thinking of Lucifer when I see him. A narrow face with fine features, high forehead wreathed by short dark hair and an almost-black goatee. He is dressed in the typical researcher's uniform – shirt and trousers in neutral shades of blue and khaki – but I can easily imagine him in an elegant suit and a silk-lined cape.

But back to what the man is saying: things are really happening. Along with his Ph.D. students Fiery Cushman and Liane Young – "some of the best I ever had" – Hauser has been studying the human moral sense for the past three years. This is a very short time as far as research goes, particularly in a new field. But in the fall of 2006, he published a book on the subject called *Moral Minds,* and a stream of professional articles has poured out from the group. The three researchers have analyzed data from over 6000 people in 120 countries throughout the world who have submitted their answers to the Moral Sense Test. I tell him that I've just taken it myself and describe some of the choices I remember.

"Straight out of the book," he says. "Take the scenarios with Denise and Frank and the runaway trolley car. The bottom line is that the two situations are exactly the same: you save five people at the cost of one. But nine out of ten people answer the way you did – that only in the first situation is it morally acceptable to pull the switch."

The dog is panting beneath the table. It has spread out over one of my feet and, even though I prod it with a toe, it doesn't move.

"The central point of our study is that moral intuition seems to be the same the world over. Quite honestly, I'm very surprised at the cultural uniformity. We have answers from men and women between the ages of thirteen and seventy; we have people who say they are very religious and people who, like you, are completely non-religious. There are people with Ph.D.s and people with only the most basic education. And none of these factors seem to have much effect on what people believe is morally defensible with respect to harming other people."

Hauser leans forward with a sly smile and I think I know what's coming.

"It's especially interesting that a religious background has no significance here. It's probably different in some European countries, but here there is a widespread assumption that morality is something that arises and persists by virtue of religion. The myth is that, without religion, we will be cast into immorality. The marriage between morality and religion is a shotgun wedding, and it's screaming for a divorce."

"And here you come demonstrating that upbringing and culture have no influence?"

"Correct."

I'm about to object and produce some counter-examples of cultural idiosyncrasies in the area of morality. What about cultures that accept honor killings of women who have brought shame to the family? What about blood vengeance as a normal method of conflict resolution? Or the Masai warriors having sanctioned intercourse with prepubescent girls?

But Hauser anticipates me and jumps ahead in the text before I even open my mouth. He tells me that the results have another interesting point. Even though there is a surprising consistency in our moral decisions, we are incapable of grounding them. We have no idea why we believe what we do. In an article in the journal *Mind and Language*,[8] Hauser and his colleagues describe in detail how they asked a group of research subjects to explain why they view a series of dilemmas differently – dilemmas formulated with variations but which fundamentally deal with the same thing.

"When we contacted people, it was pathetic what they dished out as explanations. They were completely incoherent and simply couldn't explain the reasoning behind their decisions."

Now the smile appears again.

"And, interestingly enough, this paucity of formulated principles is also true of people who have considerable familiarity with moral philosophy."

This last remark opens up a hidden door in my memory. I happen to think about the all-time superstar of moral philosophy, Immanuel Kant, and one of his more beloved formulations.

"Two things fill the mind with ever new and increasing admiration and awe, the oftener and the more steadily we reflect on them: the starry heavens above and the moral law within," he wrote in his 1788 *Critique of Practical Reason*.

But are there really laws *within* us? And if there are, where do they come from and how are they administered?

It is here that Marc Hauser proposes a new and sensational theory – namely, that we human beings are equipped with a "moral grammar," as he calls it. He draws a parallel to language and the linguist Noam Chomsky, who revolutionized linguistics in the 1950s and created a school of thought with his theory that there is a universal human grammar. Chomsky and his disciples believe that language and its fundamental structure is a faculty built into the human brain as a special module.

"The same is true of the moral sense," says Hauser. And in one of his articles, he speaks directly of a *moral organ.*[9] What he means is that, deep down, our moral positions come from unconscious and intuitive processes that take place in the brain and are based on a fundamental set of rules. A set of rules that is inborn and common to all members of *Homo sapiens.*

As Hauser catches his breath, I jump in, citing honor killings and blood vengeance as examples of moral practices that are not accepted everywhere. How does that tally with the notion of common, innate rules?

"It's quite true. There are differences in ordinary, everyday morality. But our studies indicate that there are certain classes of moral distinctions that may very well be universal."

Hauser briefly looks up at the ceiling and breathes out, as if preparing to do some heavy lifting.

"Listen. I think it's possible there is a universal set of principles and parameters that are with us from birth. These are very general, basic principles into which we can gain some insight with our dilemmas. For example, there seems to be a principle we call the intention principle – people generally always believe that, morally, the same harm is worse if it happens intentionally than if it happens as the by-product of an action with a different intention. Correspondingly, there is an action principle. A harm is more morally despicable if it happens through a positive act than from an *omission* to act. There is also a contact principle, according to which it is worse to cause harm through direct contact with the person injured than to cause the same harm indirectly. Does that make sense?"

I want to respond, but Hauser jumps ahead.

"So, we propose that there are a number of fundamental principles. But all the cultural influences we get through our upbringing come in and interact with these principles. To use an analogy with language, local culture creates English, French and Chinese morality, each with its own characteristics and peculiarities. But what we call moral grammar sets limits for what culture can do to us. There are limits to what influence from the outside can make us think is right."

Questions bubble up, and one of them is how flexible is our moral system, then? Once our upbringing has interacted with our underlying moral grammar, how stable is the result?

"We don't have the studies." Hauser shakes his head. "When I began this work, I looked around at the scientific literature and I was totally shocked to discover that there wasn't a single study on the acquisition of a foreign moral system. None! How is that possible in the world we live in?"

An excellent question.

"We need to study this. But I think this is where the analogy with language will fit like a glove. I'm convinced that moral systems work like learning a foreign language. I can tell that you probably learned English early, but if you tried to learn a foreign language now, it would be much more difficult and wouldn't succeed as well. Presumably, that's the way it is with moral systems. It's really difficult and strange to take on new moral norms, and it becomes more difficult the older you get."

At the moment, a typical reaction to Hauser's work among his fellow professionals has been to call it very interesting – in theory. Critics, on the other hand, want more substance, not just a catalogue of moral "grammar" rules but experiments that point to how those principles are encoded in brain functions. They want more on what structures we're talking about and how they operate. The fast-talking Hauser admits straight out that there is very little he has been able to offer in this regard so far.

"It's early days, alright?! We've just opened up a new field here! Chomsky started fifty years ago and, today, we still know almost nothing about the biology of language and how its structure builds on brain processes and mechanisms. So with respect to morality, of course, we know less. Yes, there is a long way to go, and when I'm dead, they'll still be working at full throttle. But just getting a description of the principles and seeing what ideas come into play – this is a huge step forward."

Ding, goes the computer. Another e-mail arrives, and the sound makes Hauser point at the machine.

"There is immense interest out there. I'm constantly getting inquiries from colleagues and to date sixty – yeah, *sixty* – students have contacted us wanting to join the group and get a Ph.D. in the subject. It's extremely hot." There is a slight pause and the hint of his previous smile. "If there's anybody who doesn't care for this development, it's the moral philosophers."

What's new is that science is pushing its way into the field of morality. Traditionally, the examination of what morality is, its results, how it arises, has been the domain of philosophers and, later, psychologists.

Typically, morality has been considered something that effectively separated people from animals, a unique human capacity. And if we look at the explanations of how human beings acquire their moral norms and positions, there have been several clues throughout time. If we go back to ancient Greece, thinkers such as Plato and Aristotle believed that the norms for right and wrong were founded in nature. Correct action was not something a human being could go out and define – rather, it was something that was in the nature of a good person. In the same unconscious way it is in a seed to develop into a plant.

Much later, we see two opposing main currents – namely, rationalism, which dominated the European continent, and sentimentalism, which for the most part kept to the British Isles. Two currents that set their cap, respectively, on sense and sensibility.

A strong exponent for the significance of sensibility for morality was the Scottish philosopher David Hume, who lived and worked in the 1700s and was particularly interested in human nature. Hume came to the conclusion that our self is nothing but a "bundle of sense impressions," and, consonant with this premise, he claimed that a moral position should be viewed in the same way as any other apprehension of a sense impression from the world around us.

For Hume, there were no universal principles – or principles at all – on which one could base morality. There were concrete situations and, when a person found himself in a particular situation, it evoked feelings and these feelings gave rise to an action or assessment. You help a person in need, because your immediate emotions command you to – helping feels good. You extend a helping hand, because you feel a strong inner prompting to do so, not because you think it is the right thing to do. If there was ultimately any reasoning or thinking involved, it was only as a later rationalization. Hume's countryman, the great economist Adam Smith, had a similar theory. Much inspired by Hume, he wrote a treatise, *The Theory of Moral Sentiments*, on how human beings need no more than emotions to judge whether something is good or bad. Rationality has nothing to offer except for considerations of what means one should use.

Diametrically opposed to sentimentalism is the super-rationalist Kant. This rule-bound German who lived his whole life in his native town of Königsberg, where citizens adjusted their clocks by his afternoon constitutionals, broke, so to speak, with everyone who came before him. Kant revolutionized moral philosophy by claiming that a moral truth cannot be derived from human emotion, from God, or from Nature. Rather, it is subordinate to the requirements of rationality just like, for example, mathematics, and, through rational thinking, it is possible to reason one's way to universal moral principles. Kant formulated the only thoroughly logical way to tackle the problem in his famous categorical imperative.

"Act only on that maxim through which you can at the same time will that it should become a universal law," it says. Or put more idiomatically – only do what you can rationally believe everyone else should also do in the same situation.

Kant's thinking has been a huge influence on moral philosophy ever since and, in modern moral psychology, it's fair to say that Kant has utterly defeated Hume and the sentimentalists. You find a powerful current of rationalism nowadays, and there are two absolute trendsetting figures: Jean Piaget and Lawrence Kohlberg. Both have formulated theories on how morality is developed in children through stages of ever-higher degrees of intellectual refinement and capacity.

Kohlberg's theory, which is based on that of his mentor Piaget and was developed during the 1960s, has been especially influential. His model proposes three general levels and a total of six stages of moral development, and it is based on the fundamental assumption that moral principles are something we pick up from the world around us – they are "active reconstructions of experience."

In the beginning, as unformed children, we learn what is acceptable and unacceptable through sheer obedience to authority. Parents indicate that a transgression has occurred by scolding or punishing a child, while praise and affection signal approval.

Later, we expand our social territory and learn from the society that surrounds us. We make our assessments based on knowledge of social conventions and the expectations of those around us, typically our families, friends and teachers. At the most developed stages, usually after puberty, moral assessments and choices come to be based on abstract thought. In the fully-formed individual, there is a consciousness of social contracts, fundamental values, and universal ethical principles. Here we see the person at his or her moral pinnacle.

Sense and Sensibility

"What we are seeing in our experiments can be described as Kant and Hume battling it out in people's heads."

Joshua Greene's office is seven floors higher than Marc Hauser's, and the view is decidedly better. Cambridge MA spreads out below us, like a plain, making me feel like a resident of Olympus. The office itself is bare and gloomy. In the middle, there are three chairs, a round table just like Hauser's, and a desk with two gigantic flat screens. Beside the door, there is a whiteboard on which the words *cause*, *desire* and *intention* are scribbled in thick green magic marker. The only visual interest comes from the huge panorama outside the window, and a bookshelf half-filled with textbooks. A wasteland of an office. I comment that it's pretty obvious that he's just moved here from Princeton, but he replies that he doesn't expect it to change much over time.

Joshua Greene is a young man with curly hair and round cheeks. The image of a story-book good boy – I imagine that as a child he was always kind and helpful, good at school and definitely his grandmother's favorite. Greene seems like a person whose equilibrium is impossible to disturb. He's also a rare bird: a classically-trained philosopher who defected to neuroscience and embraced brain scanners. He is one of the leading figures in a small and exclusive group opening up the field of moral biology, and you might say that he is one of the people who has come closest to getting his hands on the brain's moral gear box.

I need a cup of coffee and so we take the elevator down to the vending machine on the second floor. The elevators are not the fastest in the world. The trip down and up again takes forever but gives Greene the opportunity to sketch out his meandering career path.

"In fact, I can't remember a time when I wasn't interested in the big questions," he says loudly, failing to notice a sidelong glance from a girl who gets off on the fifth floor.

Little Joshua was a boy who was always asking questions – why do you do that, Daddy? How come you can do this but not that, Daddy? Daddy, tell me *why*.

Greene senior gave the precocious boy a comic book on Plato and Aristotle, and he thought it was "so cool" that there were these grown men walking around in togas, spending their whole lives asking wise questions and spouting dogmas. Twelve-year-old Joshua was a fixture at his school's philosophy debates and, when he realized that you could make a living as a professional philosopher, there was only one way to go.

"But gradually as I got more into philosophy and into the moral jigsaw puzzles, I became increasingly aware that our thinking about morality is deficient. I was most interested in right and wrong, and when you get into moral psychology, you discover that there is something strange going on. Often, you find there are circumstances surrounding moral choices that, according to all logic, *shouldn't* mean anything for the choice but do in practice."

There is, for example, the type of thing that the American philosopher Peter Unger has written about.

"Imagine you're walking by a pond where a child is drowning, and you say 'Oh, too bad, I can't do anything, because I'm wearing my expensive new Italian shoes and they would get ruined.' You'd be a monster, right? But the same isn't true if you go home and throw away a donation card to Oxfam or the Red Cross, because you'd rather use the money to get a pair of expensive Italian shoes. You can save far more children with that donation card than the one about to drown in the local pond. But refusing to make a donation doesn't make you a monster. But what's the difference? And should there be a difference?"

I reflect hazily on the stacks of charity solicitations I've thrown out over the years but don't quite reach the conclusion that I'm a monster.

"Philosophers have tried to come up with a principle that makes this distinction okay. If you are the only person who can save a drowning child, you can talk about a moral imperative. But you can modify the example a little and put several possible rescuers around the pond – that still doesn't make it okay for some of them to prioritize their shoes over the child."

This sort of hair-splitting led the student from philosophy to moral psychology. And he concluded his studies at Princeton with a thesis on how humanity can promote moral progress through a better understanding of the psychology of morality.

"You're an idealist?" I stammer, when the elevator finally deposits us on the fourteenth floor.

"Yes, but I stand by the fact that my research has a political aim. And as a student, I became more and more convinced that, if there was to be any progress, we didn't need more armchair philosophy but to understand more about how our brains actually work."

Greene pushes open his office door.

"So I went into neuroscience."

His debut there made something of a splash. In a collaboration with the gifted psychiatrist and neuroscientist Jonathan Cohen of Princeton, Joshua Greene was the first to study how moral decisions play out in the brain. This was in 2001, and the Princeton team's publication came out in the prestigious journal *Science*.[10]

"I remember the precise day I got the idea for the experiment," says Greene, setting his water bottle on the table. He puts both hands to his temples.

"It was in 1995, and I was in Israel with my parents. I was studying philosophy and had been pondering moral dilemmas for years. Then I read *Descartes' Error* by Antonio Damasio."

The famous brain researcher's international bestseller describes, among other things, a patient he calls Elliot who had a particularly interesting injury to his frontal lobes.

"He was thinking without feeling," says Greene. All Damasio's tests showed that Elliot had an above-average intelligence and was more than capable of thinking about all sorts of things and answering correctly on a variety of standardized tests. But in the real world, he made completely hopeless choices when it came to social situations. Injuries to an area of the cerebral cortex called the ventromedial prefrontal cortex had the effect of cutting out everything emotional from Elliot's life. And reason alone was not enough to navigate social intercourse successfully.

"It struck me that the difference between Elliot and normal people was almost an embodiment of the difference between Denise and Frank in the classic train dilemma!"

For Frank, it is "personal," because he has to get his hands dirty and actually push a man to his death; whereas Denise could kill "impersonally" by simply activating a switch.

"It was a real eureka moment," Greene recalls. The student vacationing with his mother and father in the balmy Middle East realized with exhilaration that patients suffering from temporal lobe injuries like poor Elliot would not see or, rather, *feel* the difference between the two situations and would answer that both were morally acceptable. But before young Greene could get out of philosophy and into proper brain research, the neurologist Mario Mendez of the University of California at Los Angeles beat him to the punch. He presented a group of patients with moral dilemmas and was able to ascertain that Greene had been correct.

"So, there were indications that emotion is necessary for certain types of moral judgments. And when I later heard about functional MRI scanning, I thought we could use the technology to see reason and emotion in action."

No sooner said than done. Greene, who was now working on his Ph.D. degree, assembled a group of research subjects who were placed in a scanner and asked to take a position on a series of classic dilemma scenarios. The researchers had divided the scenarios into "personal" and "impersonal" categories.

"You see it quite clearly, don't you?"

We're leafing through the old *Science* article, and I agree with the author that the brains scanned behaved differently in the two types of scenarios. There are some nice grey brains shown in cross-section and I count seven small red and yellow spots on them. The spots correspond to the seven areas of the cerebral cortex that showed a change in brain activity when the subjects took a moral position: parts of the brain one can, therefore, assume are involved.

"The cognitive areas," says Greene, carefully tracing with a finger the mediofrontal gyrus and two areas in the left and right parietal lobes. These are areas we know become activated when you use your working memory and typically when you undertake cool, rational analysis. These areas lit up when the impersonal dilemmas were involved. On the other hand, they are areas we know become less active when the brain processes emotional information. And this is precisely what happens when the dilemmas become personal. The light of reason is dimmed, while four other areas light up when emotional processes are involved – areas that become active when you are depressed, feel anxiety or are agitated in some way.

The conclusion that Greene and his colleagues reach is simple. If we – that is, nine out of the ten subjects in Marc Hauser's tests – believe that it is unacceptable to push a heavy-set man down on the tracks to save five people, it is because emotions come into play. Emotions that overshadow pure logic.

"It was incredible to see the results when the first images came out. Just like seeing complementary systems in the brain compete."

"A duel between sense and sensibility. Kant against Hume, as you said?"

"Hmm ..." Greene seems oblivious to his surroundings. A moment goes by before he has found another article. This one came out in the esteemed journal *Neuron,*[11] and it goes deeper into this inner duel.

"In our first experiment, we noticed that research subjects who made unconventional choices took a much longer time making their decisions. The few who answered that it was okay to push the fat guy to his death."

A time was booked for the MRI scanner to look more closely at what was going on in these strict logicians. First of all, it turned out that their long answer time went together with high activity in the anterior cingulate cortex area of the brain. It is an area that goes into action when there is an internal conflict in the brain due to reactions or information that pull in different directions.

"So, if dilemmas in which you personally have to do harm to someone are difficult and unpleasant to relate to, it's because there is a conflict between your automatic emotional reaction and a more cognitive component."

The logicians got over their initial distaste and decided to get blood on their hands in the interest of the greater good, because they *thought* more than those who declined. It turned out that they had greater activity in the cognitive brain areas that are involved in analysis and conscious control.

"If you look here at the scanner image, you can see there is particularly high activity in the dorsolateral prefrontal cortex on the right side," says Joshua Greene, pointing to the glossy page. It's sense giving sensibility a piece of its mind.

The Brain and the Ethics Committee

If you think outside the box a bit, you could say that the silent conflict taking place in our brains actually also takes place on a global plane in a far more vociferous version. It plays out in the discussions of every ethics board in the world.

Ethics has really become popular in the West over the last two decades – especially as possibilities for new, far-reaching medical treatments have appeared and biotechnology has developed in leaps and bounds. To grapple with the debate about what is permissible and what is to be forbidden, ethics boards have been established here, there and everywhere. And, everywhere, the discussion is the same. There is a fundamental clash between utilitarianism and universalism. And it is a tension that has virtually defined Western philosophy for the past two hundred years.

Utilitarianism or consequentialism, as it is also called, was formulated by Jeremy Bentham and John Stuart Mill. In essence, it holds that morality and ethics are about creating the greatest possible happiness. Or the least possible harm, depending on how you look at it. From this perspective, you can take any problem or dilemma and send it through a moral calculator to get an answer – action X provides a greater dividend of happiness than action Y; ergo, action X is the right thing to do.

Opposing the moral number crunchers are the universalists. This group builds, for example, on maxims inherited from Kant, and for them there are some universal principles and fixed, defined boundaries that may not be transgressed under any circumstances. So, it's no use to come running with numbers and calculations. A principle is a principle is a principle.

To avoid dragging out Denise and Frank once again, we can look at the dilemma for the woman in the war zone – Susan, who has to choose between killing her baby and saving the whole village or letting her child live and causing the deaths of everyone else. The strict utilitarian wouldn't hesitate to say here that it is morally acceptable – indeed, actually required – for Susan to pull herself together and strangle the crying baby. On the other hand, this solution would be condemned by the universalist, who would argue that killing is simply wrong. End of story. Even if the killing will achieve a greater good.

You can hear an echo of these arguments in the debate about embryonic stem cells that has been taking place throughout most of the world and is still going on today. It's all about the extent to which it is ethically defensible to use human embryos to create cells that can eventually replace diseased tissue and thus help various groups of patients.

Utilitarians pretty much all go along with stem cell research, because their inner calculator tells them that whatever may be negative about the sacrifice of some clumps of cells without consciousness can easily be outweighed by the positive prospect of curing people with diabetes or Parkinson's disease. However, this doesn't work for those who consider human life sacred – either on religious grounds or because they consider the embryos to be human beings and cannot reconcile using human beings as a means to an end.

The eternal war between utilitarians and universalists has

brought out the philosopher in neuroscientist Joshua Greene. He believes that you can view the tension between these two essentially different perspectives on the world as an external expression of traces left on the human brain by evolutionary development. Slightly provocatively, you can say that the universalists make use of the more primitive elements of the brain in their "no, that's just wrong" reaction; while the utilitarians light up in the more advanced and most recently arrived parts of the cerebral cortex. The prefrontal cortex, which gives us the capacity to think abstractly, exercises a higher level of cognitive control and undertakes advanced ethical calculations.

But is there a moral sense in any other creature than *Homo sapiens?* Isn't this humankind's unique invention, a characteristic that ultimately distinguishes us from other animals?

Judging by the evidence, it seems it is only a piece of our evolutionary heritage, a trait based on basic biological systems that have been passed down to us from our earliest ancestors. Thus, you can go out into the jungle and see evidence of a sense of justice – that is, an ability to distinguish between right and wrong – significantly far down the evolutionary ladder. To date, chimps and Capuchin monkeys have been observed to display something that can be interpreted as moral judgment, when they are subjected to unfair treatment.

Primatologist Frans de Waal of Emory University has worked with chimpanzees for a generation. He first became aware of the phenomenon some years ago while observing a chimp colony. In this particular colony, apes were fed once a day and they only received food when the whole group was assembled at the fence. One day, some of the young males loitered away from the fence during feeding time, and the entire group had to do without food.

The rest of the group would not tolerate this behavior, and the next day, there was hell to pay: those who had waited in vain punished the guilty parties by attacking them.

The episode got de Waal thinking, and he later proceeded to study it more systematically. In 2003,[12] he published a study involving Capuchin monkeys. These small primates were trained to hand researchers a pebble in exchange for a reward in the form of a slice of cucumber. Once they had all learned this procedure, de Waal did an experiment in which two monkeys could see each other. He gave one the usual slice of cucumber, while the neighboring monkey received a grape – considered a far greater delicacy by Capuchins. This unfair treatment made the poor monkeys who received cucumber protest loudly. Either they simply refused to cooperate and hand over their pebbles or they peevishly hurled the slice of cucumber they had received back in the face of the presumptuous scientist.

Science as Politics

"We developed in an environment that is radically different from the environment we now live in," says Joshua Greene. "And this means that we can't assume inherited rules of conduct are applicable today. The 'goal' of evolution need not be ours. Our moral sense may very well have developed, because it helps create group solidarity. But if evolution, for example, has favored groups that are good at killing other groups, does this mean, then, that modern man necessarily has to continue to do this?"

The question is raised earnestly but still sounds very rhetorical. I don't answer and, to be honest, I feel a bit thrown by its political implications as I was ready to simply move on to Greene's future research.

He is planning more studies to deepen our insights into the moral realm. Whereas he began by investigating how emotionally intuitive responses relate to more rational reasoning, he now wants to dissect the matter even further in order to tease out the neural circuits involved and understand what emotions take part and what their functional characteristics are. Greene also wants to discover what real world situations press our mental buttons to trigger the relevant feelings. Like breaking a promise, for example. With his common set of dilemmas and an fMRI scanner, he is in the process of looking at what it means to break a promise for the greater good. In the basic trolley dilemma for example, it seems to be more difficult for the average research subject to derail the train and run over the man on the branch line if she had promised him in advance she wouldn't do it. Why? Greene is zeroing in on the differences in how the brain works in the two scenarios. He is also looking for neural differences between people who decide to break the promise and those who don't.

"The *real* meaning of the sort of neuroscience I'm doing is not the actual science that comes out of it – that is, all that stuff about thinking and feeling in my experiments. No, people get really excited about imaging studies of the brain, because we're dealing with what I call *the soul's last stand*."

"You have to understand it this way: we will be able to explain an increasing number of human characteristics by physiological details and brain processes. Something like sight, for example, we have a really good explanation of. But many people think that there must be some place where the physical brain stops and something else begins. The mind or the soul. We can surrender a lot to the brain, but the soul must have some core capacities. One of them must be moral judgment, because in the Western tradition it is the quality of your soul that determines where you end up in life. So if moral judgment is not a characteristic that can be attributed to a soul, then there is no soul!"

"But that's precisely what neuroscience is doing – it's telling us loud and clear that there is no soul, we can scrap that concept."

"That's right. And philosophers of consciousness have talked about the fact that what happens in the mind is just the brain's operations seen from the inside. This materialist vision is nothing new, but it is radical; because it is one thing for philosophers to say something like this, but something else altogether for ordinary people to believe it. *Really* believe it. But when you begin to show them images of the brain with this area lighting up when you make moral choices and that area lighting up when you do something else, it gives you a very visceral experience that – yes, well, maybe the philosophers and scientists are actually right! Maybe the most intimate operations of my soul are just processes between nerve cells doing their jobs."

This is what Francis Crick called his "Astonishing Hypothesis." This venerable giant, who helped reveal the structure of DNA and touched off the genetic revolution, left molecular biology to spend the rest of his life researching consciousness. And according to his most famous statement, human beings are fundamentally "a bag of neurons."

"We are, at any rate, a bag of physical material," retorts Greene, emphatically pounding down the lid on his water bottle. He is one of the interlocutors in the ongoing debate in psychology about the extent to which it is worth doing neuroscience.

"One side is asking whether you really get something extra out of looking at the brain instead of just doing psychological studies. And they are debating what the mapping of brain functions is for."

He looks out over Cambridge. His eyes squint a bit against the reflection from the many windows of the city winking in the sun.

"When we map something, functions and characteristics, we place them in concrete places where people can see them. We change the way ordinary people think about being human. There is a cultural pay-off to what we write about in our articles – a pay-off that I believe is real and important."

Not everyone shares his point of view and certainly not in scientific circles. One example is a highly critical blog from 2005,[13] in which an anonymous neurobiologist from Maryland calling himself Lucretius takes pot shots at Greene under the heading: *Does neuroscience contradict ethics?* Lucretius scrutinizes Greene's findings on the role of the emotions and various areas of the brain, concluding that "Greene's variant of neuroscience is not science but a new extension of the category 'politics by other means'."

Politics is definitely not an everyday part of biological laboratories. Researchers typically stick to their roles as experts – nonpartisan, impartial people who provide technical advice. Greene doesn't hide his ambitions for a moment and writes openly on his homepage that he has political aims. He did not choose to study morality out of pure academic interest, but because he wants to make the world a better place. It is neuroscience for the betterment of society.

"Absolutely. But it will be in a different way than people imagine. When we think about how science helps the world, we normally think about technology. It's sort of an engineer's way of thinking that's all about building better windmills, more energy-saving cars, or medical research that can cure a bunch of diseases. No doubt, medical progress will come from my research, but I think the real benefit from this work will come in the form of what I call psychotechnology."

"Psychotechnology?"

"That is, new ways of thinking. The limitations we see in the world aren't technological in a traditional sense. We *can* actually make better cars and cure all sorts of diseases, but the solution to the big problems has to do with distributing resources and making collective decisions that benefit the world as a whole. The real barrier to progress, I think, is our own …"

"… human nature?"

"Exactly."

From his own perspective, Joshua Greene produces understanding, an understanding that must be disseminated to all people, where it can change thoughts and deeds.

"The whole *point* of my research" – and he sounds urgent here – "is that we can make the world better by understanding the biological foundation of morality. I want very much to help people develop a healthy mistrust for their moral intuition. I'm convinced that not until we understand what we are and why we do what we do will we be inclined to do things differently."

"Most of us have grown up viewing our own moral intuition as the only possible one. It's quite natural. Then, over time, we are exposed to the larger world and we discovered that there were other people with a different view of things, and we either accepted it and became a bit more cosmopolitan, or we got very upset and became sort of modern crusaders for our local worldview. That is the essence of the difference between the left and the right in the US and many other places. Imagine a world in which we grow up with an understanding that the way the brain functions was formed by random circumstances of the environment in which *Homo sapiens* developed. That human nature is full of flaws and deep, deep deficiencies and that we don't see the world the way it *is* but a filtered image of it. If this knowledge permeated culture, something would change."

Here I see my opportunity to jump into the conversation just a little, because the man who is sitting here proclaiming that the light of rationality will guide us all to a better life has just demonstrated with his excellent experiments that it is to a high degree emotions – feelings originating in primitive regions of the brain – that often govern our decisions. So, how can he trust so blindly in knowledge and understanding?

Greene halts his stream of words and actually looks a little perplexed. But only briefly, then he starts up again.

"There are two answers to that. One is that you may be right – we may be hopelessly bound to our Stone Age morality. And then what should I do with my life – go into business and make a lot of money? Or the other answer: it *may* be worth trying to save the world through science."

I can hardly say no. I don't want to appear a total cynic, with no hope for a better world and no faith in science.

"And what's more," says Greene, who has apparently come up with a third argument, "almost anything can be linked to an emotional action, if the culture sees to it. Think about drunk driving and sexual harassment!"

"What do you mean?"

"Neither of them was really thought morally reprehensible in earlier eras, right? But we've *made* them reprehensible in our part of the world. It's mortifying and there are bad feelings connected with them. In other words, there was a cultural process that transformed something accepted into something charged with negative emotions and forbidden. A phenomenon can become moralized. It happens when society recognizes that something has too great a cost and sets out to do something about it."

I'm curious as to whether all his thinking and research has affected Greene's own moral choices.

"Yeah, well, I'm certainly no saint, but there's no doubt I give more to charity than I would have, if I hadn't pondered these questions. And I think I'm more forgiving."

"A little more Zen-like?"

"At any rate, I think my behavior is more guided by the anticipated consequences than my intuitive emotional responses."

He looks out once again over the postcard panorama below and suddenly smiles. He has thought of an example, he says, and tells me a little story about how he recently went to pick up his parents at Harvard Square. He was sitting in his car in a no parking zone, when a traffic cop came by and gave him a hundred dollar ticket. Greene was irritated – a professor's salary is not that high – but, at the same time, he was able to tell himself that the poor traffic cop had a lousy job and his self-esteem depended on giving out tickets and making his job meaningful. The young Zen master would quietly hand over the hundred dollars and drive off with his family in peace.

"But then my father arrived! He was shouting and screaming about it and achieved nothing except annoyance and probably high blood pressure. There you see the difference."

Psychopaths and Euthanasia

When Marc Hauser's book *Moral Minds* came out, it was reviewed in the *New York Times* by the late Richard Rorty, a highly esteemed philosopher and professor emeritus at Stanford University. In his review, he comments sarcastically about biologists plodding into the territory of humanists and social scientists, thinking they can teach them something. People inform and revise their moral judgments by learning from history, by reading novels and philosophical treatises, says Rorty. Not by listening to biologists. And he believes it is pure hubris for Hauser to advocate so boldly for politicians and decision-makers to take our biologically-based moral intuitions into account when they form policy or make laws.

"Well, but when you ask the moral philosophers, they generally think it's a waste of time to take biology into account," responds Marc Hauser, making a sweeping motion with his hand. "For them, it's not interesting what people do, only what we *ought* to do. From my point of view, it's an idiotic attitude. However you think people ought to behave, the tenability of the behavior depends on how well it fits human nature. If as a society you want to adopt a stable moral framework – and here I'm thinking of legislation and politicians – I would prefer one that takes into consideration the knowledge we have about human nature. Then we can see where the pitfalls are."

Finally, he can't resist commenting on the presumptuous review itself.

"Rorty didn't even read the book. I wrote a response to the *New York Times*, which published it. They only rarely do this, but his review was simply incompetent."

"There is one thing that recurs again and again which people apparently have an enormously hard time grasping. I've spoken to Noam Chomsky about it, because the same thing happened with his early work on a universal grammar. It has to do with this: as soon as you utter the words person and biology in the same sentence, it turns into something fixed and inflexible in people's minds. There is nothing in my book about inflexibility, but people have a filter in their understanding that makes …"

The words run out, and you can't help but notice that this is something the biologist Hauser burns for.

"We have to get this idea out of people's heads. I mean, if evolutionary biology has taught us anything, it is this: nothing is fixed in biology. We have one genetic system – based on the same four building blocks of DNA – which has given us countless forms of life. If this is evidence of inflexibility, I don't know what ..."

Then he jumps ahead.

"You say you're interested in the impact neuroscience will have on society?"

"Yes, I …"

With an efficient movement, Hauser strokes his pointed beard a couple of times and doesn't wait for any long explanations. A lightning flash of a smile brightens his face.

"Maybe I should write about that in a *Moral Minds II*. Well, quite specifically – take the difference between action and omission."

This is one of the fundamental rules Hauser described in his moral grammar – namely, that an offender is deemed morally worse if he causes harm by a direct act than if he has done exactly the same thing by omitting to act; an asymmetry that seems quite robust across differences in culture, age, and education and, as such, appears built into our moral decision-making.

"And why is this sort of biological peculiarity so important to document, you may ask? Because I believe it has a specific meaning. Just look at euthanasia, mercy killing. In a couple of countries, Belgium and Holland, mercy killings are allowed, but in this country and many other places, you have the strange situation where active euthanasia is blocked, while passive euthanasia is accepted and supported."

He shakes his head. "Why?"

"I think that, deep down, it derives from this stupid asymmetry that tells us that acting is bad but watching passively is more okay. This distinction should have no role whatsoever with respect to euthanasia. The doctor's intent is exactly the same – namely, limiting suffering due to an incurable disease, and the consequence is the same – namely, the patient dies!"

Hauser has raised his voice and is almost shouting.

"It is actually more humane to give an active injection, because the person who is sick suffers more by waiting three days to go into a coma themselves," he says, looking actively morally indignant. It gets no better when I tell him that, at various intervals, we have had the same discussion in Denmark and that the Italians are also struggling with it. And that active mercy killing is losing in both countries.

"Grotesque! But what I want to say is that this is a case where people should realize where the asymmetry comes from, and they should be conscious of it when they ponder legal and political topics."

My smile has gradually grown a little stiff from sitting there, taking it all in, nodding and looking interested at Hauser. At the same time, it is irritating to seem so completely ignorant, so I lob the name Joshua Knobe over the net. And, luckily, Hauser returns the serve.

"Oh, yeah, that's a great example – the one with the chairman, right?"

Knobe is a philosopher but the kind who does experiments. He has shown, among other things, that the way we attribute intentions to other people is quite interesting. In one of his experiments,[14] Knobe told the research subjects a series of stories about a board chairman whose business is implementing a new policy that will improve the bottom line but, at the same time, has the potential side effect of harming or benefiting the environment.

In one version of the story, the environment is harmed, while in another, it proves to help the environment. When Knobe asked his research subjects to evaluate the chairman's intentions in both cases, he saw a clear asymmetry. Where the environment was harmed, almost nine out of ten replied that the harm was actually intended by the chairman. He did it on purpose, the bastard. In the other case, only two out of ten believed it was his intention to benefit the environment – they saw the side effects simply as a side effect. He was just lucky. So, what actually happens colors to a very high degree our assessment of other people's moral status and intentions.

"In practice, this coloring is enormously important for how you argue to a jury in a court case. If I say: this person's actions led to something despicable, many people will attribute a conscious intention to it. And in the legal system, it is crucial whether or not something was intentionally done or not. Yet another example of how knowledge of our moral make-up is important, I should think."

Something else pops into the back of my mind, and I just have to utter the word "psychopath" for Hauser to seize it from the air and without missing a beat continue his monologue.

"Yes, yes, yes. We're just working with psychopaths. An incredibly interesting group. You find a lot of them in the justice system, where they are typically charged with violent crimes. And in order to judge them, it's important to know whether they did what they did 'willfully' in the classic sense. It may be that they have no intact moral knowledge or it may be that they know perfectly well the difference between right and wrong but do the wrong thing, because their emotions are corrupted. So, it means nothing to them to make mincemeat of someone."

Hauser wants to try to clarify this issue by using his Moral Sense Test to study how psychopaths relate to the various dilemmas, where it is a matter of killing or harming others. Do they deviate from the norm and, if so, how?

"It's crucial with respect to the law whether these people *knowingly* act immorally. There have always been psychopaths who are put on trial and they are called cold-blooded and calculating and are judged accordingly, but we have no idea what comprises their mental damage. And I can give you lots of examples of patient groups in which it is important to find out what sort of neurological damage they have and the extent to which they may not know in the same way other people do what they are doing, when they do something punishable."

"You're asking what guilt actually is?"

"Yes, ultimately, it goes that deep. The challenge will be to understand what it means to do things on purpose and what asymmetries in our understanding lead to certain sorts of reactions. As far as I can see, it's all about finding the source for our moral assessments. Figuring out how factors like intention and omission come into play, to what degree we have conscious access to the principles that form the basis for what we do and to what degree training is required to undertake a given assessment."

I can sense that we're almost out of time. Marc Hauser is lightly drumming his fingertips on the table, and his gaze has wandered several times to his wristwatch. Even his old, yellow dog has started fidgeting on my foot. Finally, he has to break off for his next appointment, but he sends me off with a prophecy.

"Give it ten to twenty years," he says with his hand raised. "By that time, you'll see that knowledge about the brain will have enormous practical impact, probably in ways we can't even imagine now."

The man is right. It *is* hard to glimpse even the outline of the fifth revolution, since it has just begun to unfold. It is difficult to imagine what specific changes we are going to run into. And many of the predictions will probably go the way of many others before them – when the future finally comes, they will seem completely ridiculous. Like the happy visions in the 1950s of the distant year 2000 as a progressive age with flying cars, food in pill form, and friendly metallic robots to take care of babysitting and walking the dog. For that matter, the famous dystopias of Orwell and Huxley did not come to pass, either.

But it may not be the physical manifestation of these changes that is the most interesting. I'm convinced that the core of the fifth revolution – the really revolutionary thing – is what Joshua Greene is up to, which the venerable Francis Crick could see many years ago. Namely, that we each realize that we are and remain a bag of neurons.

It's not that I'm a stranger to the idea – in fact, I've always been a crusader for it, when the discussion comes up. In countless debates, I've found myself arguing to a stubborn dualist that "the brain is the soul," and countless times I've found myself in a heated argument advocating the abandonment of the traditional conception that human beings are divided into biology and "something else." Where in the world would this other thing come from, and where is the evidence that it exists at all? When it comes to this point, the dualist says, "We can't live with just being chemistry and molecules." To which I always answer, "Of course we can. What difference does it make?"

But this is exactly the question. What difference does it make if we seriously accept the idea that we are pure biology? What difference does it make, for example, to morality?

It is fascinating to meet someone as sure of his cause as Joshua Greene. This idealist in an otherwise tough, unsentimental field, a young researcher with bright eyes and red cheeks, replies that the result will be a greater understanding of other people. When you really *see* other people as walking complex systems, combinations of a set of inherited genes and a more or less random environment – you stop thinking about them as opponents who must be fought, because they are different. Instead, they become individuals you have to negotiate with.

That sounds good. But does it hold up?

It's probably correct, as Greene predicts, that a deeper knowledge of the mechanisms behind our moral choices and views can make a difference in the way we act. On the other hand, I believe it is terribly naive to imagine that it must necessarily make a positive difference. That the average person will feel enlightened and will automatically become more generous in everyday life, dropping more coins in the collection box when the Red Cross and Unicef come around.

But what is it that research into the biology of morality does? It points directly at the fact that there is no moral realism. The view that there must be some absolute, true values that come from nature in some way and which are independent of us is – not to mince words – a fiction. Neuroscience tells us that our sense of what is right and wrong, good and evil, comes from our own brains and is formed by the blind forces of evolution.

Of course, you can tell yourself that, if the collection box doesn't pull on your emotions or your purse strings all that much, it is because we developed in small groups, where solidarity with distant groups provided no survival advantages. Nevertheless, you can show solidarity with distant peoples through a rational decision-making process. On the other hand, you might just as well use your knowledge about moral mechanisms to liberate yourself from the "good" as cultivate it. When you know that what you *feel* from time to time is a trick evolution has played on you, you can more easily ignore these feelings.

With greater knowledge comes greater self-governance. But the other side of the coin is greater responsibility. That comes with the acknowledgement that biological traits – as Marc Hauser emphasizes – are not static but are highly dynamic and impressionable. A bag of neurons can be influenced not only from without but from within. And when you recognize that, you can no longer use biology as a crutch or excuse for your actions. You can't just lean back and excuse unacceptable behavior by saying "well, that's just the way I am," "it's human nature" or the most pathetic of all – "I'm only human."

With the fifth revolution, we're no longer "only human." We are humans with a crucial insight into ourselves, and you could say that the window into the brain that is about to be opened will give each of us moral responsibility on a previously unimagined level. Now it appears we really *have* a choice. So, whether we follow a set of culturally-fixed norms or just act on our "gut feeling" in a given situation, others will see it as an active decision.

On the whole, the issue of personal responsibility will be interesting to follow. We may very well see a development that goes in two directions at once, so to speak. On one hand, neurologists will put knowledge on the table that more or less acquits variously neurologically-challenged groups of responsibility and will have consequences for the way they are treated in the judicial system. They may be psychopaths, as Marc Hauser points out, but they may also be people suffering from some subtle damage or deviation. On the other hand, the personal responsibility of everyone else will be emphasized even more.

We can already see the signs in various debates surrounding societal issues. Global warming is not only the hottest issue around but it also illustrates how our personal responsibility stands out in the rhetoric. Supermarket products are labeled with information about their carbon footprints because it's up to *us* to make the choices that will save the planet. The drivers of gas-guzzling SUVs are the ones making the icebergs melt and the waters rise, and *they* will have to take responsibility the next time a flood hits Bangladesh and takes thousands of lives with it. Neither the task nor the responsibility can be foisted onto some convenient, shapeless collective called "society." The individual is unavoidably back in the centre.

4

HAPPINESS IS A COGNITIVE WORKOUT

My spirits aren't lifted by landing in Madison, Wisconsin. Americans call it *the heartland*, but it's cold and rainy. Plus, the airport is minuscule and utterly depressing. And there are no taxis. It all seems dismal and desolate. I go to a counter with a sign that says *Information* in large letters, but the old man sitting behind the desk does not seem eager to dispense information.

"I can't hear what you're saying," he snaps irritably, when I ask where to find my hotel in the city. He doesn't stand up, but leans further back in his chair, so that I have to lean right over the counter and repeat the name four times before he replies, only to assure me that he's never heard of it, "No such place. You're mistaken." "Thank you. I've heard so much about the hospitality of the Midwest," I say loudly, but the sarcasm has no effect on the old man. I roll my suitcase around the arrival hall, find a telephone book, and finally get the hotel – which does exist – to send their minibus.

The trip through the city is silent, but there is not much to say. There is a main thoroughfare and an endless row of family homes surrounded by lawns. My hotel, which is on the outskirts of the city and belongs to one of the big, anonymous chains, has an equally melancholy atmosphere. A noisy neon light illuminates the lobby, a large screen with a grainy picture is set on CNN, and the hotel clerk pretends not to see or hear new guests. I say "excuse me" twice, before he raises his listless gaze and checks my reservation.

"And what brings you to Madison?" he says with the voice of an automaton, as he taps my details into his computer. According to his little plastic nameplate his name is Matt.

"I'm here to learn about happiness, Matt," I say acidly. "You probably know that one of the world's leading researchers on happiness works right here at the university?"

Matt takes his fingers from the keys and looks at me for a moment, as if he suspects I'm making fun of him to his face. As if he is deciding how to put me in my place.

"Happiness research?" he drawls. Then he turns half away from me and loudly asks a pallid female lounging in a chair behind the desk, "Cherry, have you ever heard of happiness research – here in town?"

Cherry hasn't.

Matt and Cherry have not been keeping up – happiness research has been very much on the rise in recent years. More than three thousand scientific articles have been published on the subject, and a World Database of Happiness has been established, which collects and meta-analyzes happiness studies from all over the world. Recently, the field got its own *Journal of Happiness Studies*. Happiness has been a smoldering theme on the global agenda and a topic that researchers have tackled on a number of levels.

Social scientists – economists and sociologists – provide the overall perspective, when they turn their critical eye on societal happiness. A nation's overall satisfaction has become a parameter that is calculated in accordance with numerous formulae and systems. Thousands of people are questioned to assess their own happiness level on a scale of one to ten and, on this basis, regional and global happiness indices are made. The Danish media get the opportunity to remind us Danes of our good fortune several times a year, as we consistently make it into the top three on the lists of the happiest nations on earth.

The whole field arose from Easterlin's paradox. Back in the 1970s, the American economist Robert Easterlin discovered, to his own and other economists' dismay, that between 1946 and 1974 Americans had more than doubled their average income without their personal satisfaction levels changing one whit. Out of that realization came his famous article "*Does Economic Growth Improve the Human Lot?*" Ever since, a stream of articles has indicated that the answer is generally "no." Once you have achieved a reasonable income, it doesn't look like extra money makes a big difference in fundamental satisfaction. Statistically, for example, it means more for your happiness level if you are married and own a pet, while children do nothing much to increase the happiness of their parents.

Meanwhile, happiness has appeared on the political radar. In recent years, the little kingdom of Bhutan in the Himalayas has received tremendous press coverage for including a "gross national happiness" indicator as a fixed part of the national product. In Switzerland and the US, social scientists are debating how you can govern and regulate a country with the greatest possible social happiness as an indicator and, in Great Britain, the idea has reached the highest political level. David Halpern, an advisor to Tony Blair, prophesied in 2006 that, within the next decade, a government would be able to be measured by how happy it made the people. Barking at his heels was the leader of the Conservative Party, David Cameron, who stressed at the Google *Zeitgeist* conference that it was no longer enough to concentrate exclusively on the classic gross national product – one had to take "general well-being" into consideration as well. To increase the general well-being is, according to Cameron, the greatest challenge of modern society.

"A real postmodern surplus phenomenon." That is how sociologist and dean of happiness research Ruut Veenhoven from Erasmus University in Rotterdam once described happiness research to me. The Western world has solved so many of its problems that it is difficult to see what we need to do to make things even better.

Happiness is at the top of the needs hierarchy even on a personal level and we are in the middle of a crucial shift in how we look at happiness. Happiness – that transient, indefinable phenomenon – has become a good that people almost consider a right. We are *supposed* to be happy, right? Isn't that the whole point of living or what?

It's not that happiness can arise as the side effect of something else – having a family or being good at your job, for example. No, we strive for happiness in itself; we aim at a sort of distilled feeling of happiness. If you cast a glance at the swelling shelves of self-help books, they are rife with examples promising an ever-ascending spiral towards ever-higher degrees of well being, satisfaction and feelings of happiness. And whereas people once thought that it was physical health, prosperity, a good social network, a close family life and that sort of thing that led to happiness or satisfaction in life, the tone is the opposite in some of the scientific literature. Now it is being happy that leads to a better social life, better health and higher productivity. Happiness is medicine for the individual and oil for the machinery of society.

We seem to be on our way toward a society modeled on the ancient philosophy of hedonism. A society that considers it self-evident that happiness is the true end of life and in which the greatest possible personal enjoyment or satisfaction is, therefore, the individual's purpose in life. I am happy; therefore, I am.

This is a departure from the view of life that has characterized us until now. If you look at the contemporary pursuit of happiness through the eyes of previous generations, we seem to be acting like spoilt brats. Our grandparents and great-grandparents weren't happy in that sense. Or to put it another way: when they were happy, it was a side effect, it was a state that had come about while they were striving after something else. Happiness was not something you counted on or pursued.

"The intention that man should be happy is not included in the plan of Creation," said Sigmund Freud in *Civilisation and its Discontents*, which resonates well with the statement of the English philosopher John Stuart Mill that it was better to be a dissatisfied Socrates than a satisfied fool. Many would probably nod in agreement at Albert Einstein's contempt for pure happiness: "Well-being and happiness never appeared to me as an absolute aim. I am even inclined to compare such moral aims to the ambitions of a pig," the wise man once said. And you find this view again in the expression "happy as a clam," which is not so much praise as an indirect insult.

Happiness – that is, happiness for its own sake – has been considered suspect in Western culture. It is easy, superficial and unintellectual. Indeed, stupid.

But things have changed with the new hedonism and, for eggheads in psychology, psychiatry and – of course – brain research, happiness is a hot topic. My little rolling suitcase is packed with evidence. On the easily digested end, there is a tattered copy of *The Science of Happiness*, a special edition of *Time Magazine,* and *Willing Your Way to Happiness* from the *Denver Post.* On the heavy, more scientific end are provocative titles such as *The Pleasure Seekers* or, for the connoisseurs, *A Life Worth Living: Neural Correlates of Well-Being*[15] and *Well-Being and Affective Style: Neural Substrates and Biobehavioural Correlates.*[16] Hefty volumes with small print, graphs and tables.

These volumes are the evidence of a big shift in focus in the world of human emotions. For a considerable number of years, we – that is, the professionals and the rest of us – have been most interested in depression.

Depression became a hot topic when the new anti-depression drugs, SSRIs, or happy pills, to use the popular designation that every psychiatrist hates, came on the market. They opened people's eyes to the fact that quite a few people, who were not exactly paralyzed by depression but were not functioning optimally, could be treated with relatively few side effects. Depression made it on to the WHO list of international concerns, where it is now number four on the list of the most costly diseases. Everyone and his cousin were talking openly about how hard it was to be depressed. The condition, which Winston Churchill called his "black dog," was no longer a murky, taboo disturbance of the mind but came into the light as a quite ordinary disease. Almost a pandemic.

Suddenly, the bookstores were bulging with intimate descriptions of the lives of famous and creative people and their struggles with depression. The new effective pills also had their own hagiographies, of which the best known is American psychiatrist Peter Kramer's *Listening to Prozac.* On the other hand, a sort of opposition movement arose, which claimed that it was, in principle, wrong to medicate yourself into a better mood – this was something you should achieve by working on yourself in the good, old-fashioned Protestant way. Since the beginning of the 1990s, the debate has raged back and forth. Today, millions of Britons and Americans are being treated for depression, and media reports regularly raise the question of whether this might not be too much. Whether these pills are being dispensed like candy. Yet, we still hear psychiatrists say that depression is an under-appreciated brain disorder and that it is incredibly important to treat it as early as possible in order to avoid recurrence.

Now a new sort of "patient" has appeared in the treatment system. People who are *not* depressed but are demanding happiness. They cannot complain about a genuinely gloomier mood, nor are they suffering from anxiety or phobias or compulsive thoughts. There is nothing in the world wrong with them, but they march up to the psychologist for help, because they are not really happy.

Psychologists will tell you that there are more and more of these happiness hunters. They are people who come in with some vague notion that their lives should just be better in some way or other. Things are going well with work and family and all the material aspects are going swimmingly – there's equity in the villa and the summerhouse, and no tight restrictions on private consumption. But these poor people are still missing *something*.

It is something of a revolution in psychology for this lack of happiness to be deemed or discussed as a problem at all. Palpable unhappiness – that you can understand – but a deficiency in happiness? Nevertheless, it is this deficiency that has spawned one of the most booming businesses and hottest trends in psychology.

Happiness for everyone – yes, by all means, is the creed of this camp, which goes under the name of Positive Psychology. The movement started in the US some eight years ago, sparked by Professor Martin Seligman from the University of Pennsylvania. He had an otherwise respectable career studying and treating depression, but he grew tired of psychology's eternal focus on the negative, the diseased and the dysfunctional.

Classical psychology deals with making people less unhappy; it does not see its job as making someone happier. You could see this even in Freud. He believed that, if a patient finally succeeded – after intense and extensive psychoanalysis – in getting rid of his or her neuroses, the best he or she could hope for was to live with "a slight melancholy."

For Dr. Seligman that just wasn't good enough. Once it was a matter of lifting the patient's mood from minus five to zero, but now Seligman advocated that the therapist should learn how to boost the patient's mood from zero to five. Or for that matter from five to ten. And positive psychology is a success. Seligman has twenty popular books on the shelves, and *Authentic Happiness* is a solid bestseller. In the meantime, this positive awakening has spread from America and is in the process of devouring Europe, where psychological research is quietly dismantling its taboo on studying happiness.

At the moment, however, most of the research is taking place in the US, and one of the people who has worked longest on developing scientifically-tested methods for increasing happiness is psychologist Sonja Lyubomirsky of the University of California at Riverside.

"We argue that enhancing peoples' happiness levels may indeed be a worthy scientific goal, especially after their basic physical and security needs are met,"[17] she writes in a manifesto with colleagues David Schkade of the University of California, San Diego and Kennon Sheldon of the University of Missouri. The young Lyubomirsky is known for despising the pessimistic attitude of the past that an individual's happiness level is a personal trait that can't really be changed. She is trying to find the formula for a "sustainable increase in happiness."

Lyubomirsky began her career at the beginning of the 1990s, when she became particularly interested in people whom others described as really happy. Initially, she suspected that these people were happy because they compared themselves to people who were worse off than they were and, conversely, people who were unhappy invariably compared themselves to people who were better off.

The intuition of the novice researcher proved to be wrong. She found that it was indeed true that unhappy or less satisfied individuals spent hours comparing themselves to other people both above and below themselves on the happiness scale. But happy people had no idea what the psychologist was talking about – they didn't compare themselves with anyone.[18] This observation became the start of a decade's work systematically comparing happy and less happy people as a sort of window for understanding the processes at play in happiness.

It was interest from the press that made her think about making people happier – journalists who kept asking for advice to give their readers. Lyubomirsky discovered that there was no scientific literature on boosting happiness. In fact, there was a hard-nosed pessimism. Some said: it's in our genes. Others said: it's psychological adaptation. Most agreed that there wasn't anything you could do about it but accept the cards nature had dealt you. She didn't buy that dismal message and instead, began experimenting. Now she is working on a book that points out the various strategies her research has shown to be effective: gratitude, good deeds and a conscious appreciation of the good things about life.

It doesn't sound exactly enlivening to visit a nursing home or to mow your neighbor's lawn but, in controlled experiments, that sort of thing has been shown to boost happiness levels quite a bit. Similarly, Lyubomirsky has found that her research subjects feel better after they visit or call their elderly parents. She has also tried to quantify how much beneficence there has to be for an optimal effect. Her studies indicate that around five workaday altruistic acts – preferably done on the same day – can raise our feelings of satisfaction for up to a week at a time.

Or, you can make a gratitude visit, as suggested by Martin Seligman. You write a letter to a person to whom you feel gratitude and visit that person to express your gratitude face to face. A major study in which people participated over the Internet showed that, on average, this sort of activity elevates happiness levels for a month or more.

Lyubomirsky's gratitude diary uses a similar idea. Her studies indicate that you can become measurably happier by sitting down once a week and reflecting on what gave you pleasure and joy. The research subjects who did the exercise reported a significantly higher satisfaction with life; whereas a control group that did nothing felt no difference.

The Lyubomirsky group goes a step further and asks why these happiness strategies work. And some of the results indicate that it has to do with attitude. At first, Lyubomirsky believed that people who made use of her strategies would experience more positive things on the motto: smile and the whole world smiles with you. This didn't hold up. When they studied the research subjects' lives, it became clear that the individual simply began to perceive experiences, events and the vicissitudes of daily life in a more positive and satisfactory light than before.[19]

Happiness: it's to your left

So, something happens to people. The psychological level is one thing, but just as interesting is what lies a bit deeper – what happiness is inside people's heads. How does our brain create the feeling or feelings we sense as satisfaction with life? How do our grey cells decide what our mood and attitude towards existence is?

These sorts of questions are being raised a mere five minutes' walk from my bleak hotel at the University of Wisconsin, where psychiatrist Richard Davidson heads up a group of forty researchers. Davidson has been called the king of happiness research and, in 2006, his work placed him on *Time Magazine*'s list of today's hundred most influential thinkers.

Some of his most interesting – and most often cited – research has to do with the idea that our individual happiness levels may derive from asymmetries in brain activity, and in particular, how asymmetry and mood can be affected by controlled brain activity.

For more than a decade, it has been an accepted truth that happiness is almost a fixed attitude: we each have a characteristic happiness or satisfaction level that we can do nothing about. Studies have shown that we have a sort of mood *set point*, to which we quickly return after a swing in either direction. Even after great misfortunes it often takes just a few weeks before they return to the same level as before. In the same way, positive swings, such as you'd see after winning the lottery, also constitute only a temporary blip on the curve.

And when you survey the landscape, it seems there is no logical connection between personal happiness and circumstances. Compare the middle-aged perpetually happy colleague who never lets anything get him down to the beautiful young woman who is extremely gifted but who is nevertheless always dissatisfied. What exactly is going on there?

Part of the explanation, not surprisingly, may be found in inheritance – the DNA sequence you randomly receive. You can see this, for example, in a study of life satisfaction in 4000 sets of American twins by David Lykken[20] (whose name, incidentally, means "happiness" in Danish), a geneticist from the University of Minnesota. Comparisons of identical twins and fraternal twins indicated that genetic inheritance accounts for about half of the variance in happiness levels. While genetics determines a whopping fifty percent, factors such as income, religion, marital status and education altogether only contribute a measly eight percent. The rest, argued Lykken, is purely up to chance and is something you owe to the ordinary vicissitudes of life. His conclusion is bleak: "It may very well be that any attempt to become happier is as futile as the attempt to become taller."

But pinning it to DNA does not provide an answer. Genes are just strings of information that hang out in cell nuclei; they don't do any work in the organism. If you want to get a better answer, you have to ask what happens in the brain, and this is where Richard Davidson comes in with fMRI scanning and EEG measurements. Put somewhat simplistically, he discovered that the primary seat for happiness or satisfaction is the left prefrontal cortex. Or rather, the basic mood of the individual, the set point, appears to be a reflection of the relationship between the level of activity in the left and the right sides of the brain – in the sense that a higher relative activity on the left side than the right provides a higher level of happiness.

Richard Davidson and a team of colleagues demonstrated this in a major study from 2004[21] in which the researchers recorded EEG signals from eighty-four women and men, whom they also asked to answer a series of questions. What was their current mood, their general well-being, and how were things going with such psychological parameters as self-confidence and personal development? There was a clear connection between the ratio of electrical activity in the right and left cortex, on the one hand, and the person's basic mood or happiness level, on the other. The more the balance was skewed to the left, the higher the happiness level.

An EEG uses electrodes on the scalp to register the overall electrical discharges in the cerebral cortex below, and it does not provide fantastic resolution. As long as people speak in general terms such as prefrontal cortex, that's fine, but you cannot exactly pin down from which three-dimensional area the activity is coming. You get excellent spatial resolution, however, with a functional MRI, and Davidson did studies in which he found that current moods went together with characteristic patterns of activity. When you are emotionally stressed – worried, angry or depressed – there seems to be activity in a circuit that includes the amygdala and the right prefrontal cortex. If the person in the scanner, however, reports being in a good mood, the activity in these two areas is lower, while correspondingly there is more activity in the left prefrontal cortex.

At lectures and conferences, Davidson is known for trotting out a very characteristic bell curve. It is based on the scans of up to 200 research subjects in "resting state," that is, not affected by any particularly negative or positive stimuli. The points on the curve show the relationship between the individual's activity in the right and left prefrontal cortex. And it seems that you can scan any random person and, according to where they wind up on the curve, you can predict the person's general level of happiness. The further to the left on the curve they are, the happier and more satisfied with life – the further out to the right, the more dissatisfied and the greater the risk for real depression.

Before coming to Madison, I had written ahead and offered myself as a guinea pig. Of course, I know myself to be quite a disgruntled and even surly individual at times, but it would still be interesting to see a proper, objective assessment. Unfortunately, there is a polite refusal in my inbox from Davidson's secretary, Susan Jensen, who informs me that the scanner is completely booked while I'm in Madison. Ms. Jensen has spoken to the responsible technician who says it is impossible to arrange an extra trip through. In fact, there are no happiness scans taking place at all that week.

Audience

"Welcome to the Waisman Center. Richard Davidson's group? Yes, that's on the second floor. You can take the elevator or the stairs over there, honey."

Honey? When a receptionist calls you that, you know you're out in the country. And you can't get around the fact that Madison is miles from the nearest real city. It's located right in the middle of a state known for its cows and dairy products. But the university does not suffer for that. The University of Wisconsin, Madison has an excellent academic reputation, which has lured the top researchers in the field, and is one of the top five universities in America when it comes to raking in the most research grants.

You can actually see the money. At this end of the campus, where the medical and biology departments lean up against the large hospital, there are cranes at work everywhere. You stroll past huge holes in the ground, where new buildings are shooting up, and the "old" ones look like they were just put up yesterday.

The Waisman Center is a beautifully-appointed building with high ceilings, huge windows, blond wood panels everywhere, and something indefinably streamlined, which gives it a business-like character. Up on the second floor is the Laboratory for Affective Neuroscience, where the tone seems to be pure, effective professionalism.

"I've made a chart of all your appointments," says secretary Susan Jensen, handing me a print-out of a spreadsheet with names and times: 11:00 to 12:00, meeting with Dr. Richard Davidson. 12:00 to 1:30, break. 1:30 to 2:00, meeting with Dr. Catherine Norris. 2:10 to 2:30, observe scanning. I am twenty minutes too early according to the program and anticipate that I can take advantage of the time to amble about and soak up the atmosphere, perhaps even chat with someone.

"If you could just wait here on the sofa, I'll bring them out to you."

"But can't I . . ."

"Coffee?" Susan makes me a cappuccino from a spanking new machine and makes sure I sit down in a deep leather sofa in the large common area. Stranded here, I can sit and observe a lawn and a narrow strip of parking lot.

Susan eventually comes out of her office to report that my meeting with Davidson has to be delayed fifteen minutes and, unfortunately, they can't extend it on the other end, because of another appointment.

That I have come all the way from Copenhagen via Boston and Sudbury apparently means nothing here. After entertaining myself by observing the parked cars, I kill some time studying a long frieze covered with photocopies of various media pieces on Waisman Center researchers since the mid-seventies. They are arranged in chronological order and all the clippings are laminated in clear plastic. Towards the end, close to our own time, there is an intriguing, almost bizarre picture. It shows an Asian man, a Buddhist monk, sitting in the lotus position with bright orange robes wrapped around his body like petals and with a halo of electrodes around his head. Behind him, you can glimpse some Westerners, busy with wires and machines. Richard Davidson is one of them.

"He's on the stairs." Susan waves me toward her boss' office. Just outside, there is a white board, where someone has drawn a classic Thai Buddha, with tight, stylized curls and a calm smile. The first thing you notice in the office is a photo of Davidson bowing respectfully before the Dalai Lama, who smiles in the same way as the Buddha but is wearing horn-rimmed glasses. Apart from the little photo, the office is empty, bright, almost anonymous.

Richard Davidson looks like the pictures I've seen in *Time Magazine*. A small, thin man with a narrow face and thick, wavy hair, which was once entirely dark but is now mixed with grey. A brown tweed jacket and a green woolen pullover make him look like a professor from Oxford or Cambridge. But, really, what is most striking is his total equanimity. There is something unflappable about the man, which makes me feel flustered and insecure. Suddenly, my tongue curls up and, as I try to introduce myself and explain what I'm doing there, all I can think about is the embarrassing pronunciation mistakes I'm making. He listens with a slight smile that might be welcoming but is also infinitely distant.

"I've been studying the brain for thirty years and, of course, been interested in positive feelings, but the people who support research are primarily interested in negative emotions. There is a connection to diseases such as depression and anxiety, so with respect to funding, the negative was always the lab's bread and butter."

Then he gives the slightest smile.

"But it was clear early on that our capacity for maintaining a positive emotional state is not just interesting in itself; it governs how vulnerable people are to depression and anxiety. Your capacity to sustain positive emotions is an important ingredient for your resiliency."

I simply nod and am told that the idea of looking at the remarkable asymmetry with respect to happiness levels is actually quite old. It derives from observations made in the 1970s of depressed people with brain injuries – patients who suffered a blood clot in the brain or a cerebral hemorrhage that only affected one cerebral hemisphere.

"We could see that patients with injuries in regions in front of the left cortex and in the deeper-lying areas called the basal ganglia were far more vulnerable to a subsequent depression than others. This observation was important. It indicated very directly that the left prefrontal cortex plays a critical role for positive emotion. When this particular brain circuitry is disturbed or destroyed, that person's vulnerability to depression will be increased, because they are incapable of sustaining positive emotion. Remember that one of the primary symptoms of depression is that it is impossible to experience joy, no matter what happens. You just lose interest in things and don't take on goal-oriented activities. Goal-directed behavior, having an appetite for life and being extroverted – these are all functions or activities that are controlled or carried out by the left frontal cortex."

Davidson next brought in some research subjects to play with their emotions. They were shown pictures or read things that were supposed to make them sad or upset, and the researchers could see that the moods affected the prefrontal asymmetry. But it was apparent fairly quickly that the variance between individuals was much greater than the difference between positive and negative emotion in the individual.

"They come in with their set point – what we call their happiness level. When we began to study it thoroughly, we discovered that the level we can measure in the frontal regions is stable over time in adult individuals. So, if I tested you today and in six months, you would show the same pattern."

I get a tremendous desire to bring up the scanner again, despite the rejection I got by e-mail. After all, it came from a secretary. And I really want to know where I am on the scale – how *far* down the unhappy right branch of the curve.

"So, this genetic set point ..." I begin but am interrupted.

"Listen, we have to be careful and not necessarily attribute the happiness level to genes."

"Yes, but don't your own experiments with crying babies say something about it being inherent?"

I have just read about Davidson's spectacular experiments measuring the frontal brain activity in infants of six to ten months of age.[22] While these toddlers were with their mothers the researchers measured the electrical activity in their left cortex; then they asked the mother to leave the room. Some of the children screamed like stuck pigs at the separation, while the rest took it stoically and played on, undisturbed. Davidson was able to accurately predict from the measured brain activity which children would break out into howls.

"We tested infants in more experiments later and were able to ascertain that the activity measurements you get only predict their behavior in the relevant situation. As opposed to adults, the measurement does not predict behavior or resistance to stress in the long run. You just don't get the same measurements over time in children."

"They still have a chance to push their set point upwards?"

"There is, at any rate, much more flexibility early in life. From our assembled data, we believe there is great plasticity in the system until around puberty and then the system's set point or base level becomes more rigid. Even though … we can come back to that, but let me quickly stress that it is not *fixed.* You brought up heredity, and there is certainly genetic influence, but I believe that environment far overshadows the genetic contribution."

"David Lykken's studies show that it's sort of fifty-fifty."

"Those studies, in my opinion, don't hold up. I'm familiar with more recent evidence indicating that there is at best thirty percent heredity in these traits."

Okay, okay, I'm not going to use my allotted time to argue about numbers. So I ask whether he has any ideas about the mechanisms behind our individual happiness level.

"It's still an area that is being researched intensely. But let me make one thing clear. Even this – that there is a set point for something like humor and mood – is by no means strange. It is a characteristic we know from pretty much all biological systems. Your respiration and heart rate have a set point, your temperature regulation has a set point, a value they oscillate around. And when we talk about brain processes, which are complex, I think there are a lot of forces working to stabilize the system."

He places both hands up to form a pyramid. "There are probably good evolutionary reasons for there to be stability in our mood."

"Why?"

"Because ... well, we know that mood influences our view of our surroundings, our attentiveness and our cognition, our ability to analyze and think. So if there was a huge variation in mood, you could imagine that it would disturb our interaction with the environment."

Of course, it doesn't sound expedient in a survival context. But there *are* mood swings.

"Purely personally ...," I say and am about to recount some interesting details about what it's like to experience quick and powerful mood swings. But I don't get it out.

"Yes, thanks. I wasn't finished. What I wanted to say was that everyone has a set point, but there are huge individual differences in mood dynamics. So we have to distinguish between mood and emotions. Some people have great emotional range but are rock solid in their mood, that is, their set point. They can shift emotionally up and down very quickly, while their basic mood remains the same over time. Then there are other people whose mood is unstable and who may be irritable and sad for five days and then change."

"Yes, I myself . . ."

"I was about to say that the Dalai Lama is a good example of a person who has a very stable mood but whose dynamic emotional range is enormous. He can show enormous changes in emotion very quickly, but he is always happy."

Before he loses himself in personal recollections, Davidson comes back to his set point and stresses that it is not actually set for life.

"We know that the measurable activity patterns can be changed and that's where our focus lies today."

It is here, too, that the Tibetan Buddhists come into the picture. It turns out that they are world champions at being "happy," at least according to EEG measurements and MRI scans. There is also an indication that their significant capacity for happiness is not inborn but is the result of meditation.

"They tell us, in other words, that a directed mental activity can affect the whole mental apparatus," says Davidson.

It is a bit strange to sit here across from a well-groomed, tweed-clad, top academic, and imagine him meditating for forty-five minutes every morning. As a young Ph.D. student at Harvard in the 1970s, Davidson went to India and got his first taste of intense Buddhist meditation, and he has practiced it every day since. Nevertheless, he did not come out of the closet until five years ago, after "having kept it under his hat."

"In my personal life, it has been every important. But there was no room for meditation and a career in neuroresearch, so I didn't speak about it to anyone. Everything changed when I met the Dalai Lama for the first time in 1992. I discovered that it was time to come out of the closet and speak about it publicly, also because I could play a catalytic role, since I had scientific credibility from my long career. And it has been very satisfying to see the positive reaction from even the most hard-nosed scientific colleagues."

It's true, there has been a response. Meditation is becoming more and more popular and, at the moment, is being studied by research groups all over the place. Princeton's Jonathan Cohen – morality researcher Joshua Greene's old partner – is looking into how meditation sharpens awareness and, at the University of California in San Francisco, psychiatrist Margaret Kemeny is studying the extent to which schoolteachers can increase their powers of empathy through meditation. To the north in Canada, psychologist Zindel Segal of the University of Toronto has found that meditation is an effective way to prevent a relapse in patients who have had several clinical depressions.

The Dalai Lama himself is extremely interested in neuroscience. In his book *The Art of Happiness*, he anchors the effect of his meditation practice in the brain's physiology. He writes that "the systematic training of the mind – the cultivation of happiness, the genuine inner transformation by deliberately selecting and focusing on positive mental states and challenging negative mental states – is possible because of the very structure and function of the brain."

At the turn of the millennium, he gathered a group of the most prominent researchers in the field to a dialogue in the Indian city of Dharamsala. In 2005, he was invited to be the main speaker at the annual conference of the Society for Neuroscience – an event that brings together some twenty thousand international researchers. The year before, he wrote an opinion piece for the *New York Times* challenging researchers to study whether there were elements of Buddhist practice that could be "translated" into the secular world and do some good out there. Today, he continues to dispatch his best lamas – the "Olympic athletes of meditation," as Davidson calls them – to chilly Wisconsin to take part in experiments.

Here, the professor emphasizes that "it's not about Buddhism as such. It's about investigating how mental processes can modulate states of mind. And these monks do something special – they cultivate introspection and develop things like empathy and positive affect. Happiness, if you will."

Researchers hope these mental athletes will provide a way of understanding how the emotions are expressed in the brain. One thing that leaps out at a brain researcher is that the EEGs of these practiced monks show a powerfully increased gamma rhythm[23] as soon as they begin to meditate. The gamma rhythm consists of certain electrical charges or brain waves, and the more powerful they are, the more precisely the network of brain cells communicates back and forth. You can compare it to a jazz band that sounds so much better when they are synchronized than when they each play to maximize their own sound. Davidson has an idea that gamma rhythm may be a signature for the way the brain can change itself. And his people have data to indicate that the more powerful the gamma signal is, the more intense the subjective meditation experience.

The monks have also taken a trip through the MRI scanner – both at rest and during various types of meditation. And generally it's the case that they far exceed normal research subjects when it comes to left prefrontal activity. Researchers have seen some of the highest levels in monks at home in the Himalayas who spend the majority of their lifetime meditating.

But there is hope for mere mortals yet – as is indicated by an experiment in which employees from an American biotech company – all strangers to meditation – were invited to an eight-week course on "mindfulness meditation."[24] The technique, which is a sort of non-religious filtrate of Buddhist traditions, has to do with monitoring or being attentive to one's thoughts and feelings, focusing on the positive ones and dropping those that threaten discomfort. Mindfulness meditation has gradually become very fashionable at hospitals and clinics in the US as a means for reducing stress and increasing well-being. And just like positive psychology, it's creeping across the Atlantic and establishing itself in Europe.

One of the experts is Jon Kabat-Zinn of the University of Massachusetts Medical School. He and Davidson enlisted a team of high-gear, stressed-out employees from the biotechnology company Promega to take a two-month meditation course, three hours a week, finishing up with six days of intensive meditation training. A control group from the same company that received no training was also measured. Everyone received an EEG of their frontal lobe activity and was questioned thoroughly about their general state of being.

At the outset, pretty much everyone complained of stress and worry, and many admitted that they weren't doing well at work. After eight weeks, the answers were different. In general, the stress level was down amongst those who meditated; their worries did not fill their lives any longer, and most reported feeling more energetic. They even felt a positive desire to work.

The researchers' measurements also revealed that changes had occurred inside their heads. The notorious relationship between left and right prefrontal cortex activity had moved clearly toward the left.

"My hypothesis is that a strengthening of neuronal circuitries in the left prefrontal cortex occurs," says Davidson. He believes that it all has to do with circuitries that in some way inhibit signals that would otherwise come into the region from the deeper-lying amygdala, a structure that is involved and active in the processing of a number of negative emotions, particularly anxiety.

"But we didn't just look at effects in the brain. The group that meditated also showed a measurable effect in the immune system, which had become more robust. After an ordinary flu vaccination, they showed greater reaction and created far more antibodies."

In fact, it was the case that the more active the left prefrontal cortex had become, the more powerful was the generation of antibodies. The study immediately became a media darling, and American reporters hailed a milestone and a breakthrough. Hype that Richard Davidson does not much care for.

"It was an interesting result," he asserts dryly. "Because it showed that regular meditation for just two months can lead to noticeable changes in brain circuitry that seems to control our basic mood. But we still don't know how much these circuits can change and how long the changes last. In the Promega people, we could still see the changes four months after the course, but we didn't follow them longer than that."

"But some of them, so far as I've heard, carried on with a mindfulness practice."

"That's correct."

"Is it conceivable that you can only permanently change your happiness level by working on it permanently?"

Richard Davidson nods slowly and swivels in his chair a bit.

"I think so, and I want to compare it to physical training. If people exercise, they get into the shape they want to be in, but if they stop, most of the effects of their training disappear. I think it will turn out to be the same with cultivating positive emotions. You can see it with musicians, right? Even top, world-class musicians continue to practice; they can never escape it. I think that a crucial part of the future will be a recognition that a phenomenon like happiness is not a transient feeling or a fixed set point, but an *ability*. An ability that can be trained and developed and shaped."

"Now, you said a little while ago that our set point looks like it is set sometime around puberty. Shouldn't you be interested in childhood – because that's when you can really accomplish something?"

"It's a huge part of our thinking right now."

Suddenly, there's a different kind of enthusiasm in his voice, and a couple of times Davidson even rubs his hands together spontaneously. He explains that a research and education initiative has been launched to set an agenda for studying school children from first through twelfth grade. It is an ambitious program that is supposed to figure out what specific activities you can give children of different ages and how you can measure the effects – both what happens in their behavior and what changes happen in the brain.

"It hasn't been published yet, but we have preliminary data indicating that just five minutes of directed activity a day can produce long-term changes. That is, if they're introduced at the right age and carried out consistently. We have collected data from training where children cultivate compassion and generate a feeling of benevolence and kindness. We try to cultivate empathy for others – again, an ability that I think can be trained to a certain degree."

I can almost hear an echo of Marc Hauser back in Cambridge, his point that the new biological knowledge tells us that we can *change* our biology. Gone are the days of being able to say "that's just the way I am."

"I agree." Davidson straightens up a bit in his chair. "We are on our way toward discovering that our personality in all possible dimensions is far more malleable than we thought. And it's going to give us a more fluid concept of the self and thus a different attitude toward changing our way of being. But let me say that I've been out talking to laypeople about this over the past couple of years, and there is real resonance. People are motivated to enter into this new way of thinking."

Thinking over Feeling

"Isn't he just impressive?" Catherine Norris has large, bright eyes and appears to be full of benevolence and amiability. "I think he handles things incredibly well. Just think – here's a man who is one of the top people in emotion research, he has a group of forty people to look after, he is always incredibly busy, and yet he never seems rushed. Could you sense his calm?"

"Absolutely." I nod. "It must be all that meditation."

Catherine ignores my sarcasm. The young psychologist from Chicago has been a member of Davidson's group for two years, but she has apparently never quite gotten beyond being impressed.

"You know he's one of the people who attracts the most money to the university?"

I didn't know that, but on my tour of his little kingdom I can see that it lacks for nothing. Charming offices and large, well-lit, well-equipped laboratories. There isn't anyone in the EEG lab – "it's usually always busy" – and the sophisticated machinery is quiet. No computer noise, the keyboards are silent, and some weird-looking helmets with 128 electrodes, each with accompanying cord, hang loosely on the wall. They look almost like living creatures – sad, mutated jellyfish that have been left on shore, just waiting for the tide to reclaim them.

"I can see from the schedule that you are going to see a scanning and then meet Tom Johnstone," says Catherine. We compare spreadsheets to be completely sure, and it's correct. So we head down to the basement, where the center's various scanners are set up.

Inside, in the long corridor, the atmosphere is the same. I automatically lower my voice and start to tiptoe, so my shoes won't make any unnecessary noise. We creep along the walls and look cautiously in at the little PET scanner, which is for research animals only. A technician shows us a little contraption for holding rats. A bit further along, we come to a suite, where there is a sign reading: "Keep out, MRI scan in progress."

"If you write about this, you can't use her real name," says Melissa Rosenkranz about the day's research subject. The young researcher is about to investigate how negative feelings affect asthma symptoms, and the asthmatic girl on her way into the scanner seems nervous. She sniffles and gasps a little. I promise to call her Alma, if I write anything. As I take my position in the back of the control room, she is placed on the scanner's narrow stretcher and is given a red panic button, before she moves into the central cylinder. She looks like a skinny hotdog in an oversized bun.

"If you're comfortable, we will turn off the microphone and run the first scan. It will take about seven minutes. And remember to lie completely still," says Michael, the technician, from behind the pane between the scanner and the control room. He starts up the machine, and it sounds like a jackhammer, as it takes one picture every second. On a monitor on the other end of the control room's large console, you can see a grotesquely enlarged image of one of Alma's eyes. The image is being transmitted live from a small camera in the scanner and fills almost the entire screen. The eye blinks and moves in all directions and, because the screen has a sort of network of lines at cross angles, it looks like a tiny animal in a cage.

Soon Alma's cranium and its contents are flickering on Michael's large screen.

"Look at that brain – it's extremely wide. I need to make some adjustments," he says. The young man, who runs the scanner eight hours a day, sees a bit of everything thanks to his job. Today, he can show everyone how a severe sinus inflammation looks under magnetic resonance. Everyone in the room giggles a little and looks closer at the shapeless grey shadow behind Alma's nose.

"Mucus," explains the man standing beside me in a white surgical gown. He collaborates with the brain researchers and is a doctor at the university hospital's lung department.

"But a little secretion doesn't mean anything," he assures me. Lots of times, there is something interestingly abnormal in the brain itself – there's pretty much no such thing as a normal brain.

"It can look scary. Strange hollows, shrinkage or awkward anatomy, but for the most part those sorts of irregularities don't mean anything for the person. Not so long ago, we had a research subject who had a sizeable growth, the size of a lemon, in one hemisphere. He had no idea and wasn't told."

My disbelieving look makes the nice doctor lift both hands to ward off objections and he provides a more detailed explanation.

"Mind you, there's nothing strange about that. Everyone signs an informed consent form, where they check off whether they want information about abnormalities without any practical significance and whether they want to know if there is any suspicion of something serious. From time to time, we find something that looks serious. Malignant tumors. So we send them to radiology for a diagnosis."

He exhales demonstratively. "It's not pleasant to have to confront a person who comes in as a volunteer to help us with some research and goes home as a cancer patient."

We shudder together. And suddenly I'm glad Davidson wasn't able to arrange one of his scans for me.

The researchers finish with Alma and go to their respective offices, and I once again find myself on the black leather sofa, lost in the view over the lawn.

"If the weather wasn't so grey, the foliage out there would be quite nice," someone says with a flat Australian accent. The man who suddenly appears beside me looks a bit like an Aussie rugby player with a solid build and unkempt, mousy hair. "Tom Johnstone," he says, thrusting out his hand. He plops down on the sofa, where I'm sitting staring emptily at the distant trees. I tell him what I think about the American Midwest and Johnstone is sympathetic. He and his scientist wife would much rather be living in a charming European metropolis with cafes and nightlife and culture, but in their world, research is the determinant.

"And you can do first-class work here in Madison," he says, shrugging his shoulders.

Johnstone's into some very interesting things: "Cognitive strategies for modulating emotion." It's all about how we can affect or regulate our emotions with conscious processes – good, old-fashioned thinking, if you will. This undeniably lines up with the new focus that happiness can be created and modified from within.

"Over the past few decades, there has been great interest in emotions, but everyone has focused their gaze on the role feelings have at the cognitive level. We realized that emotion affects our choices to a significant degree and thus our analyses of information, of situations and of the world around us."

That's all fine, says Johnstone, waving his hand a bit lazily. He, of course, agrees that these are important points.

"But there has been a tendency to look exclusively at everything as *bottom up*, you know. That is, there were some deeper-lying emotional regions in the brain that sent messages up to the higher cognitive areas."

Of course, I know where he's coming from. I am familiar with the concept of the primitive lizard brain situated down in the unconscious, full of murky, ungovernable feelings that bubble up and contaminate cool logic. Moreover, the trend is to "listen to our feelings," presumably because they are thought to hold some deep, but hidden, wisdom.

In psychology, however, people have been talking about the phenomenon of emotional reappraisal. That we human beings can inspect our feelings and, through this inspection and evaluation, temper them. Tone down anxiety or anger by telling ourselves to pull things together. In other words, cognition that affects emotion.

"*Top down*, as we say," concludes Johnstone, slapping an article down on the table. It is quite new, I can see, and the title alone takes up three lines.[25] I flip through it quickly and remark that there aren't many pictures. You get to page five before you find a series of MRI images of brains seen from above and from the side. They are not much bigger than postage stamps.

"No, but the text is quite interesting in its way, if you take the time to read it," Johnstone responds.

Roughly speaking, the results state that the relationship between cognition and emotion looks like an internal tug-of-war, where the emotional end yields when you pull on the cognitive end. The experiment itself is simple enough. Normal research subjects are put in the scanner, where they are shown a film with negative emotional content. Among them, a clip of a tennis player losing the Wimbledon final with its accompanying signs of grief and sorrow: shouting and screaming, hanging of the head, throwing of the racket. Then the research subject is asked to feel the pain of this poor millionaire professional tennis player.

"It initiates activity in several areas of the limbic system, a part of the brain whose primary role is to process and expedite emotional information."

So far, so good. The next step was to ask the emotional research subjects to dial down the intensity a notch and actively dampen their negative emotions.

"Here, you can see that the limbic activity actually drops, especially here in the amygdala, which is definitely implicated in negative emotion." Johnstone points at one of the postage-stamp pictures. "But this suppression occurs in the context of the person showing increased activity here" – he finds another postage stamp – "in areas in the prefrontal cortex. It is the ventromedial prefrontal cortex that is activated in both hemispheres."

This region is typically involved in cognitive processes and the more activity the person presents here, the greater the suppression in the person's amygdala, and the greater the success the person had in toning down the emotion. But not only that. The researchers also measured the level of the stress hormone cortisol in the subjects over the space of a week, and they could see that those who most effectively regulated their brain activity also had the lowest and least volatile hormone concentrations.

"A healthier cortisol profile," says Tom Johnstone, just so we're on the same page. And if this is the case, it is presumably directly due to the fact that the amygdala is being suppressed. One of the amygdala's tasks is to activate the body's stress response by sending messages to the hypothalamus, which in turn passes the message to the adrenal glands that they should inject cortisol into the bloodstream.

"It struck me how great a difference there is in how good people were at this regulation. It will be incredibly interesting to find out what does it, and now that we've established a basic understanding of the participating brain network, we can move on to look at other factors."

He examines me through the top part of his glasses.

"Are you interested in neuroeconomics? You should be. Last week, we had a visit from Ernst Fehr, who is an economist from Zurich and one of the people studying how the brain functions in choice-making. And he believes that people who have a better ability to regulate their emotions are actually better at making choices."

MODERN LOTUS-EATERS

Later, on my way through the campus back to the city after my meeting with the happiness researchers, I feel decidedly exhilarated. At one point, a construction worker almost runs me down with his fast-moving forklift, but I just shrug my shoulders and shoot him an affable smile. It is the research that does it. Not so much the actual results, which are still very limited, but the new focus on exploring how our cognition can be made into a mental resource. "It opens the door to a new freedom," I can't help murmuring to myself. A field of research that reveals how conscious ideas can be harnessed to give shape to unarticulated feelings and urges is a powerful tool for coping with life, delivered right to your door. This way of thinking really grabs hold of me, because it is in line with my own personal experience with happiness.

Or, rather, unhappiness, because what I went through was a depression. Nothing unusual, just an ordinary collapse, the kind millions of people experience. But, still, a collapse that changed my outlook by driving home a decisive point.

I found myself on an emotional rollercoaster, where things quickly spiraled out of control. By the end what I was experiencing was not a bad mood or a bad case of ennui but rather a complete abandonment to hopelessness. I was barely able to get up in the morning and drag myself through the workday, and only turned in my assignments to keep from losing myself entirely. As soon as I walked through my own door, I fell apart and was no longer a person but one big, blistering sore. I could cry for hours and when total exhaustion finally sapped the air from my lungs, I would collapse and do nothing but sit and stare into space.

There was nothing but a strange pain that was both physical and yet something more, which I blamed on unhappy circumstances and the vicissitudes of life. *Of course*, I was unhappy, I told myself. How could it be otherwise? Here I was – a woman in her mid-thirties, with a professional life that was a catastrophe and a private life that was a fiasco. A hermit with few friends, who had absolutely no reason to care about anything in particular.

Self-hatred and self-accusation became a nerve-wracking chorus, the only thoughts that circled through my mind. Finally, it became too much for my family. Against my will, they dragged me to the doctor, who didn't need to ask many questions before she uttered the word "depression" and prescribed an SSRI compound.

I went home and dutifully swallowed the antidepressants without any great faith in their effectiveness. Mine was an existential pain caused by a life in ruins! Not more than a week went by, when the raw pain seemed to have dampened, as if packed in a layer of isolating cotton wool. After another couple of weeks, the darkness was dissipated by what was unquestionably a feeling of happiness. I was through with crying and could stand in the middle of the street – even on a rainy day – and feel elated just to be alive. There was absolutely no external reason for things to have taken this turn – not the least change in my life or external circumstances – it was all down to chemicals.

Those little white pills proved that the emotional pain I had experienced had a chemical and physiological content and that it *didn't* come from the outside. And it could be changed from within. The "self" that saw itself as particularly ill-suited for life and viewed the world as an evil stepmother was just one chemical version of myself. It was, so to speak, a special filter through which I perceived reality.

Actually, I had known this already. As a seasoned science reporter, a biologist and even ex-neuroscientist, I was not unfamiliar with the facts. But it was only now that the purely intellectual knowledge was transformed into a subjective experience that the message finally sank in.

Which is not to say that I was cured of all bad moods – that would be impossible – and I can still experience episodes of real depression. But I've installed a brake system to make sure I avoid a descending spiral of thoughts that ends in the muck. A *cognitive* brake. In the form of the simple acknowledgment that all of my feelings and moods are, in the end, "just chemistry." Pain, jealousy, anger and hopelessness are ultimately just particular patterns of activity in the labyrinth of my brain. No more, no less.

This is a powerful acknowledgement when it sinks in, because it allows you to step back and observe yourself with a cool and analytical pair of eyes. It's like standing in a circus ring with a roaring tiger and suddenly discovering that the animal obeys commands with a light swish of a whip. With a little ingenuity, you can even hop on and ride the tiger.

And if you imagine this sort of knowledge becoming widespread in society, we are back to the fifth revolution. Think about it. When it becomes a matter of course and common childhood wisdom that happiness or satisfaction is a chemical condition – literally, a *state of mind* – there must be consequences. You could, among other things, imagine that this deeper self-knowledge would help slow down the chase for everything we *think* is happiness but actually does not make us happy. All the things that one study after another indicates have no lasting effect. Status symbols, designer kitchens, SUVs.

Of course, we know this already – it is no accident that the saying "money can't buy you happiness" recurs in every religion and philosophy. If we nevertheless chase after the material, it's because our emotional tiger is biologically programmed to line its pockets. This raw *desire* for more things, higher status, always pulls us in. But the question is whether insight into our internal mechanics will help strengthen the cognitive self and make it more capable of achieving what Biblical parables and Buddhist teachings have not.

Another aspect of this insight is the possibility of a technological fix. Could there be a shortcut, without the cognitive mental workout, in the form of medicated happiness?

In the long run, yes. Of course, the current antidepressants do not increase happiness levels in healthy people; but it is clear that, if happiness is fundamentally chemical, it can also be manipulated chemically. Happiness specialists also agree that, in *principle*, it must be possible to alter your basic happiness level purely medicinally.

"With sufficient knowledge of the neurotransmitters and brain areas involved," as Richard Davidson said before my audience was over. True to his own research, he stressed that cognitive behavior strategies have a more penetrating effect than today's medicines. But, in the end, we're talking about the same thing – namely, a chemical effect on the brain, and Davidson predicted that the future will offer more targeted and effective drugs. Correspondingly, positive psychology's guru, Martin Seligman, has argued that you can easily medicate yourself into an immediate feeling of joy, happiness and satisfaction. He also presumes that you will be able to help people to the state of self-forgetfulness he calls *eudaimonia*, which he believes is an element in any proper, solid happiness.

Time will show the extent to which he is right – or, perhaps, rather when he will be proved to be right. And while the drug industry is working on it – and we can be sure that they are – an increasing awareness that we are in fact "just" chemistry will eventually deflate the on-going debate about the extent to which happiness via prescribed chemicals really *is* happiness.

As it stands, for the many people who are prescribed antidepressants, this sort of chemical happiness is viewed as "artificial" and, in some way, inferior to the "real" version. With a proper understanding of the brain, however, this sort of foggy thinking falls flat. The same thing would, without doubt, happen to the modern argument that life is supposed to contain a certain quantity of unhappiness, sorrow and travail to be complete; an argument which largely derives from the completely unfounded belief that unhappiness and pain are in themselves cathartic experiences that ennoble the mind. Happily, you don't have to suffer to be an artist – if the results from the current happiness research indicate anything, it is that creativity, productivity and ordinary extrovertedness all rise along with happiness levels.

But the flagellants can rest easy for the time being, for no one will ever be able to go through life without feeling at least some form of unhappiness. Disease, death, and unrequited love will always be with us; unemployment and homelessness will undoubtedly endure; and natural catastrophes will continue to strike. We will never master such things. But while brain research will not protect us from all evils it is on its way to providing us with better weapons to fight our battles with.

5

OUR SOCIAL MIND – IT'S ALL ABOUT SIMULATION

In his play *No Exit*, Jean-Paul Sartre lets one of his characters conclude that "Hell is other people." And in some senses, he is right. If it weren't for all the restrictions other people put on us, life would be easy. Instead, from cradle to grave, we are effectively hemmed in by the actions, demands and rules of others. Some of our earliest conscious experiences have to do with parents or other adults making us do things we don't want to do. Brush your teeth, take a bath, wear your nice, plaid skirt to kindergarten. In school, we are powerless, all our time and activities defined by teachers. And, as adults, when we think we can finally decide for ourselves, we learn the hard lesson that our freedom is very effectively restricted by other people – colleagues, bosses, tax collectors and, especially, spouses and offspring.

Even though they can be hell sometimes, "other people" are the closest thing we can come to what might be called heaven. The same lovers, children and friends who make demands on our time and invade our personal space also give us the experiences that make life worth living. We only experience really powerful, deep emotions in interactions with other people. It is no accident that isolation is the worst punishment to be doled out in penal systems that don't endorse direct, physical torture. Because, since we are incurably social creatures, it *is* torture to be without other people.

We evolved from a line of apes and hominids that, for millions of years, lived in small, tight groups or large families. And, according to one classic theory, their social habits were one of the driving forces in the development of the large brains and high intelligence we boast today. The line of thought is simple. Intelligence is constantly developed and increased because it provides an advantage for doing well in a changing social context. You can think of it as a furious race to be the best at reading others. The better you are, the more difficult it is for the competition to get away with cheating or deceiving you, and the greater your own chance of cheating and deceiving others. Thus, the lie is born. And researchers who know their apes have dubbed *Homo sapiens* the Machiavellian primate.

This doesn't sound like praise, but social intelligence has received a lot of notice. Daniel Goleman's book *Social Intelligence*, for example, was an international bestseller. Goleman examines the advantages that socially-gifted people have – advantages that don't seem to correspond to their raw, clinical IQ (which has been the usual measure for intelligence). Social intelligence for Goleman is the ability to empathize with other people and make them feel comfortable and, ultimately, to get social situations to work. Such traits obviously pay off in a world in which more and more is taking place in loose, dynamic social networks instead of strictly-defined hierarchies.

Neuropeople have also become more interested in social contexts and social skills. They talk about a social neuroscience or a social cognitive science, and it is a field that is experiencing huge growth. In the last five years, it has developed from nothing to a cutting-edge discipline that appeals both to young, talented researchers and to those who dole out research grants. A handful of specialized journals has hit the university libraries; conferences have been arranged; associations are getting on their feet.

It all revolves around perceiving the brain as a social organ. As individuals, we're woven into all sorts of social webs and contexts and our brains are ultimately designed for interaction with other brains. With the ambition of bridging the gap between the biological being and the social being, researchers began to study how the nervous system itself is involved in social transactions. In practice, this means grappling with phenomena such as prejudice, personal attitudes, and social conflicts in relation to the neuronal and physiological processes they are linked to.

Prejudice provides a good example of how this new research is proceeding. With piles of studies backing them up, social psychologists have ascertained that we all suffer from "prejudice" in the sense that we have a tendency to put other people into categories – such as "us" and "them" or "good" and "bad." These mental categories provide a sort of built-in bias that is automatically triggered when we meet people who fit them. A number of psychological studies have shown that we meet pretty much all other ethnic, religious and political groups with prejudice. Despite political correctness and every possible good intention, prejudice is almost impossible to eradicate. But where does it come from? Or, rather, how is it encoded?

Psychologist Mahzarin Banaji of Harvard and neuroscientist Elizabeth Phelps of New York University studied this a few years ago in an experiment involving black and white Americans. The researchers put their subjects into an MRI scanner and showed them pictures of faces. The faces were all those of strangers, but some were white and some were black, and the color difference clearly prompted activity in the amygdala. One purpose of this small region is to signal when there is something emotionally meaningful afoot, and, as expected, it was activated by all the unfamiliar faces. When they were shown several times, however, the activity in the amygdala went down but only in connection with faces that belonged to the research subject's own race. When a black person saw a white person, and vice versa, there continued to be activity in the amygdala.[26] In fact, it was revealed that the more prejudiced the person proved to be in psychological tests, the greater the activity.

If these automatic prejudices "live" in the amygdala, it appears that political correctness belongs in the frontal cortex. In a follow-up study, Banaji and her colleagues called in a team of white research subjects to whom they showed pictures of black and white faces. The pictures were visible for 30 milliseconds at a time, and there was the usual pattern of increased amygdala activity, when the black faces were shown. On the other hand, when they were shown for 525 milliseconds, something interesting began to happen. The amygdala activity went down as regions of the frontal cortex began to be activated – regions associated with evaluation, regulation and control.[27] The greater the frontal activity, the more powerful the repression of the amygdala.

Research teams have applied the same process to other phenomena with a social meaning, which has led to a series of studies that shows, for example, where things happen in the brain when we exercise self-control, or when we regulate our negative emotional impulses. It may sound crude or even futile to focus on active brain regions. The high-tech scanning exercises have somewhat pejoratively been called "light-bright" psychology. However, the end goal is not to discover where in the brain stereotypes or self-control are expressed but to illuminate which cognitive and emotional processes create complex social behavior. This knowledge must then be taken back to social psychology, where it can be used to identify how a given action on a social level is to be explained.

Another current running through all of social neuroscience – and social psychology for that matter – is an interest in some of the basic social skills that characterize human beings. There are things we master to a degree no other species can approach. We can put ourselves in someone else's shoes. We can learn from others just by watching or listening to them. And, in particular, we can acquire culture – which is to say, assume very different ways of acting and thinking. These abilities work superbly together and, in reality, constitute biology's most cunning invention – namely, a liberation from biology's own limitations. For what happens with the invention of culture? Our behavior is no longer strictly defined by genetically-determined instincts and coded patterns of behavior but can suddenly bloom in all directions. We have become inventors of behavior. And with our inventions, we can develop independently of our inherited biology and radically decouple ourselves from the raw genetic evolutionary process.

We're talking about a quantum leap. And for the world of brain research, the big question, of course, is what it is in our brains that makes this leap and these marvelous human abilities possible. One serious theory – and one of the breakthroughs of recent years – is the so-called mirror neuron system. Mirror neurons, put simply, are brain cells that allow us to mirror others. Or you could say they are brain cells that have two, almost mirrored, functions. On the one hand, they participate actively, when we undertake a particular action – typically, a movement – and, on the other hand, they are also active when we see someone else do the same action. You can immediately see how such a two-in-one mechanism forms the basis for a sort of internal simulator. Moreover, this very simple simulator can be a generator of complex social characteristics.

"I predict that mirror neurons will do for psychology what DNA did for biology," neurologist Viljanur Ramachandran has said. And his parallel is based on the fact that mirror neurons provide an overall framework for exploring and understanding a whole host of our mental abilities – abilities that until now have been virtually inaccessible to experiments.

Mirror neurons have provided fuel for new experiments, and via mirror neurons researchers are happily attacking such complicated matters as empathy, language, and imitation. But the beginnings were more humble. These fascinating cells were originally discovered in monkeys, and it happened through one of those random chances that come up in research. In 1996, at the University of Parma, Giacomo Rizzolatti and his colleagues Leonardo Fogassi and Vittorio Gallese had been studying how rhesus monkeys control their hand movements. Researchers planted delicate electrodes in individual neurons in particular areas in the monkeys' cerebral cortices – namely, the so-called premotor cortex, whose specialty is to represent and govern movement. Quite specifically, they were concentrating on an area known as F5.

The electrodes coming out of the cortex were connected to an amplifier in the laboratory, and every time one of the observed cells became active and fired off an electrical signal, there was a loud noise like someone snapping a finger in front of a microphone. The expert research monkeys had already been trained to reach for peanuts and other food, so the researchers could study the behavior of particular neurons in relation to particular movements. They measured what happened when the monkeys picked something up from a flat surface, when they reached for something, and when they picked something up and simply held it.

The experiments went according to plan, and their observations came rolling in. But during a break something unexpected happened. The rhesus monkeys had no more food to reach for and were thus unoccupied, while the researchers were taking a break and had left their equipment for a moment. Suddenly, the amplifier squawked. The sounds indicated that the monkey brains had begun to signal as if the animals were reaching for something. When Rizzolatti looked over, the animals were completely quiet with their hands in their laps.

However, his colleague Fogassi had just reached over to a fruit bowl to take a banana. At first, no one understood what was going on. They thought the equipment was playing tricks or there was some sort of technical glitch. But there was nothing wrong with either the electrodes or the amplifier and, when the researchers began to systematically repeat the movement towards the fruit bowl, it never failed to trigger the monkeys' premotor neurons. This was clearly no accident – they had happened upon a real phenomenon.

The finding was promptly reported and, while colleagues the world over discussed its implications, the group in Parma threw themselves eagerly into exploring the new phenomenon more closely – to lay a solid groundwork. They discovered that there are also mirror neurons in other areas of the brain – namely, parts of the parietal lobes, which also deal with movement. And they concluded that only a small number of all the neurons in these areas have mirror traits.

The next logical – and really interesting – question was, of course, whether the mirror system also exists in humans and how it functions. Now, as most people would object to wiring sprouting from their scalps, the electrodes had to be abandoned for indirect methods for measuring brain activity. Rizzolatti and his colleagues themselves conducted the first studies with PET scanning, and other groups tackled the problem with MRI and EEG equipment.

In the beginning, as with the monkeys, they worked with movement and measured their research subjects as they reached for objects or observed others reaching for something. And they saw the same pattern as in the primates. One of the first things they discovered was activity in the areas of the frontal cortex that correspond to the monkeys' F5 and in us goes under the name Broca's area. With such indirect measurement, of course, you can't say for sure that it is the same cells that become active during observation and deed, but the activity in the two situations converges so much that it is the natural conclusion. And today the consensus is that there are systems of mirror neurons in people and that they exist in both the frontal and parietal lobes.

This hasn't made researchers give up on monkeys. In Parma, the rhesus monkeys haven't had a free moment since the initial breakthrough. The work has led to the discovery of a broad spectrum of mirror neurons that act according to their specialized function. Hence, in the parietal regions, there are mirror neurons that, "know" the difference between when a seized object is to be eaten or just put somewhere. Some only fire when another monkey takes a piece of fruit and puts it to its mouth, while others are active when the piece of fruit is placed in a container. Similarly, Rizzolatti and his colleagues have found that there are special classes of mouth mirror neurons in the monkey's cerebral cortex that deal exclusively with mouth movements. Some react to the sort of movements involved with eating, while others are activated by "speech" movements.

Speech, of course, has special relevance for *Homo sapiens*. And theories have been proffered that mirror neurons may well have played a role in our development of language. Broca's area, where we have a system of mirror neurons installed, is one of the central regions for language processing. And the neurons here are activated, for example, when you move your mouth and hands or see others do so. Marco Iacoboni and his group at the University of California at Los Angeles have connected mirror neurons even closer to the use of speech. They have shown that just listening to someone speaking triggers the areas of your own frontal cortex which are normally active when you speak yourself.[28]

Not only that. Iacoboni has also discovered that we, apparently, run an inner simulation of the scenarios described to us. This came out in an experiment in which a group of research subjects was read sentences that described a particular action and they promptly reacted with activity in parts of the premotor cortex that otherwise govern the same actions. Simulation thus promotes understanding. But more recent experiments indicate that simulation is an absolute necessity for us even to be *able* to understand. In an as-yet-unpublished study, the UCLA researchers used strong magnetic stimulation to put the active mirror neuron areas out of play while the reading was going on. The result was that the research subjects no longer understood what they were hearing. The sounds went through just fine, but they were no longer comprehended, because the internal simulator wasn't working.[29]

Simulation presumably plays a crucial role in our impressively well-developed ability to imitate. We see or hear someone do something, and this directly activates some of the brain cells that are needed to execute the relevant actions. But identification with others' experiences also seems to be within the work remit of our highly advanced mirror neurons. At any rate, this is the message of some sensational experiments done by Christian Keysers at the University of Groningen in the Netherlands on the sense of touch.[30] Keysers brought in a group of volunteers, popped them into his MRI scanner, and had a research assistant brush their legs ever so lightly with a feather duster. As expected, this tickle triggered brain activity in the secondary somatosensory cortex, where this sort of sensory input is registered and processed. What was surprising was that the same brain activity was triggered when the subjects in the scanner were shown a brief video in which an actor's legs were tickled with a feather duster.

When Keysers announced his findings at a conference, he showed his research colleagues a clip from the classic James Bond film *Dr. No* in which Bond wakes up with a hairy tarantula the size of a hand creeping up his arm. And as Keysers explained, the fact that we can "feel" something when we see it happen to someone else comes down to strategically placed mirror neurons.

The enterprising Dutchman has also looked at the feelings of disgust and loathing. A group of research subjects were made to smell rotten eggs and rancid butter while they were lying in the scanner and, just as expected, there was activity in their insula, an area of the cerebral cortex that is consistently activated when we feel disgust for something. The same insula also lit up when the research subjects watched a video clip of a man smelling a glass filled with liquid while making a grimace of sheer disgust. As the title of the article for the journal, *Neuron,* says: "We both feel disgust in my insula."

"I see mirror neurons as the entryway to an existential neuroscience," says Marco Iacoboni, when I first call him. Iacoboni, who heads UCLA's Ahmanson-Lovelace Brain Mapping Center, immediately agrees to a meeting, and he even offers to arrange a little scanning experiment with me as the subject. "If you're interested, that is."

When I got the offer, I was of course extremely interested. But since then, I have been to Madison, witnessed an experiment, and heard the scientists talk about how many seemingly healthy volunteers emerge from the machine with a head full of abnormalities. So, now as I stand outside the squat Ahmanson-Lovelace building and press the door buzzer, I'm not quite so keen.

It is empty beyond the doors in the central atrium, with neither patients nor research subjects – except for me. Marco Iacoboni comes out of his corner office to say hello. The first thing I notice is his relaxed manner and European style. The slender man with the combed-back silver hair has lived for many years in the US and made his career here, but he has preserved a touch of an Italian accent with its characteristic twang. And there is something pleasantly laid-back about him. Not a trace of the nervous energy and suppressed aggression that is so typically American. Iacoboni asks with interest how I'm doing, how my trip was and, when he smiles, there is an amiability that reaches all the way to his eyes.

"Have you been scanned before? Well, then, let's get started. I just have to find Jonas," he says, leaving me on the sofa of the waiting room. The professor himself doesn't dabble in the technical details of the experiment – that is the job of his colleague, psychologist Jonas Kaplan, who comes down from the second floor. He is slim and dark-haired, and seems awfully young to me. "I've put hundreds of people through the scanner," he says by way of greeting, as if he could read my mind. Then he gives me an informed consent form to fill out; a long questionnaire to absolve the researchers if I walk away with unintended effects from the scanning.

I solemnly swear that nothing is wrong with me, that I'm not being medicated, that I don't have any metal in my body – no screws in my hips or pacemaker in my heart. The one filling I have in my tooth from my school days doesn't count, fortunately. Finally, the form explains that I cannot expect any personal gain from participating in the present experiment. On the other hand, it informs me that society as a whole may benefit from the new knowledge acquired about the functions of the brain, knowledge that may help us to understand the brain and thus treat its diseases. With a good conscience, I sign.

"If you're finished, you can take those earrings off," says Kaplan, offering a little box. To be on the safe side, I also whip the ring off my finger and am ready to encounter the magnet.

"I'm sorry it's so cold, but it's necessary for the scanner."

The skinny young man shivers when he escorts me into the large room with bare concrete walls.

"I just don't understand why it has to be so damn cold in the control room, where the rest of us work," he mumbles. With goose bumps all over my body, I lie down on the thin stretcher that will take me into the scanner's narrow cylinder. It doesn't look so bad when you're on the outside looking at other people go in, but up close it's very different.

"There's not much room in here, is there?" I say without

getting any reaction. But, of course, I've already ticked the box on the form to tell them that I don't suffer from claustrophobia.

"Cram these in good. It's pretty noisy," says Kaplan, handing me some blue foam earplugs, which I work as far into my ears as possible. Then he gives me a couple of earphones to top them off and, finally, a thick pair of what look like diver's goggles that are going to run a film for me. They are taped securely to my head.

"Here is your panic button. If you start feeling bad, just push it."

The little rubber ball is connected to an alarm in the control room, and I clasp it to my chest, while Kaplan stuffs pillows on both sides of my head to immobilize it. I sense that I'm slowly sliding in but, fortunately, can't see the walls of the cylinder two inches from the tip of my nose. In the goggles, there is a slide show with panoramas from exotic beaches that remind me of the winter catalogue of a charter tour company; white sand and small, thatched beach cabanas, where drinks are being served. There is also a series taken under water, of angelfish swimming around above a starfish-studded ocean floor. On the sound front, there is a strange, deep, soft thumping like a regular heartbeat. It's the scanner idling.

"Can you hear me? Are you okay?" I answer yes to both questions and try to lie still. I can hear myself snort when I breathe, and I try to keep my head still when I swallow.

"We're just scanning your brain to get a quick structure. It will take about fifteen seconds."

Then, a sound starts up, like someone using a jackhammer just behind my head. When it stops, a sign appears in the goggles – *Are you ready?* It says. I am. Then it displays a cross, which Kaplan tells me to focus on.

"Here comes the film. Just lie still and watch it."

At first, there is a neutral blue background with a cup. Then a hand comes in from the right and grabs the cup. The next picture shows a fat little teapot and a plate with cookies, with the same cup in the middle of the tableau. Then, the hand slowly comes in from the right and takes the cup. I wonder a little about what's going on and speculate on what they are able to read from this. Then, the scene changes, and the cup is in the middle of a sea of crumbs and crumpled napkins. The hand returns to get the cup. Again and again, this bizarre sequence is shown. It seems like an old surrealist film or an early attempt at making video art. Finally, it stops, and we cut back to the angelfish and the blue sea.

"Now we're going to do an anatomical scan to get the structure of your brain in high resolution. It will take seven minutes."

This time, the sound is different, consisting of a loud click every other second, when the machine takes a picture. Finally, it stops, and Jonas Kaplan comes in and takes me out of the cylinder.

Angelfish are still swimming before my eyes when I enter the control room, where it *is* pretty damn cold. Jonas Kaplan is sitting in front of one of the huge computer screens, and he pulls up some gray scanning images. A couple show a cross-section from the bridge of the nose to the neck, with the eyes sitting like large round balls at one end. Just like headlights. Suddenly, I recognize my own profile in longitudinal section from the top of my head down through the whole cranium, and the first thing that hits me is the makings of a double chin. I can only hope Kaplan hasn't noticed. My nose – which I never liked – seems excruciatingly large and the overall impression seems sort of bunny-like, since you can see my front teeth with roots up into the jaw.

"So, here's your marvelous brain. That's *you*," says Kaplan, smiling cheerfully. Finally, I focus on the beautiful, whitish folds inside the cranium and think back to the brain bank in Boston. There is the same strange feeling of much too much intimacy. Then, the anxiety from before suddenly returns.

"Is it normal?" I stammer.

"Completely. At any rate, I can't see anything abnormal."

The proverbial weight falls from my chest. No tumors, no ominous brain shrinkage. I send a wordless thank you out into empty space. On the other hand – "normal" doesn't sound especially positive. It has the ring of "ordinary." I stare once again at my gray matter.

"Isn't there anything at all out of the ordinary?"

Kaplan raises his eyebrows and looks at the images again, lightly shaking his head.

"No, nothing immediately comes to mind."

"It's not particularly large or, maybe, particularly small? Nothing in the folds or details that looks special?"

"Sorry," he says with a shrug but then thinks of something.

"You have a very cute corpus callosum. Look here," he says, tracing its elegant curve with the tip of his pen. I have a feeling he's just trying to reassure me, but actually, it looks cute to me too, and quite strong.

"There is a huge variation," says Marco Iacoboni, who has suddenly and soundlessly appeared in the room. "We see people with the IQ of Einstein, whose corpus callosum is quite thin."

I don't quite know how to take that piece of information but don't have a chance to answer before he continues.

"You should be glad there's nothing suspicious, because that's often synonymous with problems. A couple of weeks ago, we had a young girl in who was apparently completely healthy and walked out with images of an advanced brain tumor."

"Yeah, that wasn't fun," remarks Jonas Kaplan.

It takes time to transform the raw data that comes out of the scanner into the beautiful color images we are familiar with as a map of brain activity. You have to make computer models, which are fed magnetic signals, and you have to correct for the subject's small movements and other quirks.

"Because I'm Italian, everybody thinks I was at Rizzolatti's lab for the original discoveries," begins Iacoboni as we start to look over the results of the scan. He tells me he has only met his famous countryman at conferences around the world.

"But when I spoke to him for the first time in the 1990s, he was interested in using scanning techniques to study these new cells and, when I heard about them, I was sold. I mean, human beings have unique abilities – we can talk, read, express complicated emotion, and have intentions – but we come from organisms that have nothing but simple motor systems. Suddenly, we find mirror neurons, which indicate an elegant way of making the leap."

Iacoboni grasps the front of his shirt like a romantic tenor. "I just knew that I had to be part of this research." Then we look at my data.

"We've run the experiment before, and it indicates that mirror neuron systems are part of unscrambling what other people are intending to do in the future."

It turns out that the results have already been published[31] on the basis of twenty-three other volunteers who observed the blue background, the tea cup, and the hand coming in from the side.

"You show quite typical activity," says Iacoboni, pointing to a bright orange spot a bit above my right eye. It shows a "very robust" activation of my right ventral premotor cortex, and things seem to go as expected every time the hand reaches into the picture.

"But the amazing thing is that there is a difference in the activity in the various situations. In this image, the table is clean, right, while it's full of garbage in the other. So, the context of the situation indicates that the hand is doing something different – that it's going to take the cup to drink from it or to clean up, respectively. And it turns out that this area on the right side becomes much more active when the context is getting ready to drink something. At the same time, this area shows more activity in both contexts compared to the clean picture, where the hand is reaching in over the blue background."

I notice that the last image he points to doesn't have much orange.

"Look, this says something about the fact that mirror neurons aren't just interested in the action they're viewing but what motivation or intention the action expresses. That's fantastic! Being able to dissect the details of how our brain decodes these sorts of subtle details. Wow."

Iacoboni has worked himself up, and I agree with him that it is absolutely fantastic. But does this mean that mirror neurons will be neuroscience's answer to DNA?

"Oh, that's Ramachandran's expression. I love it! With just a few words, he has effectively branded the entire field, hasn't he? But you were asking why research into mirror neurons should mean so much. I can answer that on several levels. On the academic level, people have believed up to this point that our brains work with sharply divided systems that sense and comprehend on one side and act on the other."

Iacoboni takes a small paperweight and a CD and puts them beside each other on the table between us. He points to the space between them.

"Everything that goes on between sensation and action should, according to the old model, be cognitive processes of different sorts. But that is a hopeless way of doing business. The brain has evolved to do things as quickly as possible, and the quickest way is to endow the system with the simultaneous ability to act and to comprehend. This is precisely what mirror neurons do."

He places the paperweight on top of the CD.

"But okay, this is, as I said, academic. Mirror neurons are important on a different level. For they show – all the way down to the neuronal level – how we are connected. There is a physical connection between you and me. When I do something here, the same thing goes on inside you; you simulate me, so to speak, in your mirror neurons."

Now we're dealing with what Professor Iacoboni calls existential neuroscience. Then I ask him what this really means and he breaks up in laughter.

"If only I knew! No, kidding aside, I admit it sounds hazy and is difficult to explain, but I believe we are dealing with a revolution. Brain research has been driven by a strict analytical tradition, but mirror neurons have exploded the traditional notion of thinking about separate systems delivering information to each other in a linear way. There is no assembly line in there. Instead, we've got a system of brain cells that process my actions and your actions in a holistic way in one and the same procedure. They simulate and re-create not only others' actions but also their intentions and feelings. Just being in the world means, on a fundamentally physiological level, being directly involved with others. Just think about that," Iacoboni points back and forth between us. "Our nervous systems are directly connected."

The deepest connection between one person and another, of course, is in empathy, where not only do you know what the other person is feeling but feel it with that person and feel a need to act on that person's behalf. Empathy is a huge theme in the young science of mirror neurons. And Marco Iacoboni helped kick-start things with an article in PNAS in 2003, in which he scanned subjects who were shown a series of photographs of faces with diverse emotionally-charged expressions of anger, fear, joy or disgust. At first, they just observed. Then, they were asked to imitate the relevant expression themselves. And this showed that there were areas in the frontal cortex that were active in both situations.

"There are, in other words, mirror neurons involved when we express ourselves and when we comprehend the expression of others. We say that mirror neurons fundamentally help us understand the actions of others, generally speaking. But some of these actions express the emotions of others."

But how do you get from comprehending these emotions to actually *feeling* them? Everything seems to indicate that it's through strategic communication links. Those brain regions that belong to the so-called limbic system and which are involved in the processing of emotions do not contain, so far as we know, mirror neurons themselves. But the UCLA researchers studied the anatomy. And they found communication links that went from the frontal areas that are active in connection with imitation to areas in the limbic system.

"So we're proposing a model in which mirror neurons first

comprehend the actions of others and, at the same time, simulate these actions by becoming active. But in connection with this activation, they send signals to emotional areas, and it is through this communication that you really *understand* the feelings of others. Again, what we have is a little internal simulator. The decoding is not, as such, cognitive simulation, where we make an effort to figure something out. It is an immediate, automatic and unreflective form of mirroring."

I think of the psychopaths Marc Hauser mentioned. What has been said of them is that they don't feel any immediate empathy but are good at reading and figuring out other people. So what happens with their mirror neurons?

Iacoboni squints a bit, when he admits that this is beyond his field of expertise.

"Oh, I would love to study psychopaths, but it's a difficult group, a bit slippery, hard to get a handle on. And where do you get hold of a team of volunteer psychopaths?"

That question is allowed to float unanswered between us, until Iacoboni picks up the thread.

"From other groups, we know that there is a connection between the degree of empathy an individual can muster and the activity they have in their mirror neurons."

This holds true, for example, in a group of ten-year-old children with whom the researchers have experimented. The children, who were easy enough to get hold of, were given an ordinary psychological empathy test that, among other things, involved looking at pictures of faces and decoding the frame of mind they express. Later, they were put in the scanner to have the activity in their frontal mirror neuron areas measured while they looked at the same faces and imitated their expression.

"We saw a clear correlation. The higher they scored on the empathy test, the greater the activity in their mirror neuron system. And I believe that you can consider the system as a biomarker for empathetic ability."

"But this ability, is it fixed or can it be modulated in some way?"

"That is what is really interesting, isn't it? And there are data that indicate this. I can show you what we've seen in autistic children," says Iacoboni, who proceeds to print out a copy of an article that supports a hypothesis that came out around the turn of the millennium, claiming that the development of autism has to do with a breakdown in the mirror neuron systems. There are many degrees of autism, but at the core of every case is the fact that the emotions are affected. Autistic people have a notoriously difficult time reading and understanding the feelings of other people, and this defect makes it difficult for them to cope in social situations. They are, in the words of author Temple Grandin, herself a sufferer of autism, like "anthropologists on Mars".

At any rate, they have problems with their mirror neurons. Along with psychiatrist Mirella Dapretto and several others, Iacoboni studied ten high-functioning autistic children and compared their reactions to facial expressions to normal peers with the same IQ. And whereas normal children have significant activity in the mirror neuron areas, the autistic children showed no activity. In the classic emotional areas, they proved to have far lower activity than normal children.

"Here, too, we can see that the more serious their autism is, the lower the activity," says Iacoboni. "But back to your question about modulation. We know that, if children are purposefully trained to show an emotional message in their facial expression, they become better at doing it. Of course, we haven't had time to see whether this coincides with greater activity in mirror neurons, but I think this will prove to be the case."

He also believes that autistic children can teach us something about how the mirror neuron system develops.

"There are autistic children who typically refuse to look at others or make eye contact, while they are small, and I think this prevents them from training their mirror neuron system. The interplay between mother and child is fundamental, and it presumably goes both ways. Little baby smiles, then mommy smiles back, and baby learns in his mirror neuron system to connect the mirror neurons to a smile and to an understanding of the emotions that the smile reflects."

Suddenly, Marco Iacoboni packs up his own smile and becomes serious.

"Listen. If we understand this incredible system better, it will in the long run provide obvious opportunities to train empathy and thus social competence. Imagine how much you could achieve by using this in a school context."

Iacoboni inhales audibly and stretches his arms out over the table.

"On the whole, I think that neuroscience needs to get out into the social sphere, out into society. The notion that the brain can be formed and influenced is incredibly important. I mean, there is a neuroscience revolution going on, and the whole idea behind this must be to create *better* people."

Only a moment goes by before he smiles wryly and looks down.

"Yeah, it doesn't sound good to say it like that, but I really believe it's all about being able to direct and guide human nature."

THE OLD MAN AND THE BRAIN

It looks undeniably like a trend. All these top researchers are saying that their data and results need to get out into the real world and have some effect on general society. There is no academic fear of intimacy here, no stuffy air of the ivory tower, and this is a long way from the behavior we normally see from biologists. The usual routine is to stress that they are only scientists, professional experts, employed to produce knowledge; what this knowledge is used for is for other people to say.

It's as if I'm hearing the echo of a point made by neurologist Antonio Damasio back in 1994 in his classic *Descartes' Error*. He writes that "knowledge in general and neurobiological knowledge in particular have a role to play in human destiny." And he continues in this grandiose vein, saying that a deeper insight into the brain and the mind is the way to achieve the happiness and freedom for which we irrepressibly strive and which have been the driving forces behind the constant struggle for progress in recent centuries.

These are big words. But it looks as if the sixty-two-year-old Damasio himself is now taking stock of them. At any rate, he has ostentatiously thrown himself into social neuroscience as the head of the newly-created Brain and Creativity Institute at the University of Southern California, which is just a half-hour drive from Iacoboni's base.

You might say that Antonio Damasio helped break the ground for the current interest in social neuroscience. The famous neurologist and neuroscientist hails from Portugal, but from 1976 to 2005 he worked at the University of Iowa, where he made his brain research department the center of the world for exploring the role of the emotions in our mental life. And he deserves a great deal of the credit for today's recognition of the high degree to which people navigate the world with emotions as the motor and the intellect dragging along behind.

Damasio has always worked in close collaboration with his wife Hanna, who is considered one of the world's leading experts in imaging techniques. And, together, throughout the 1980s, they concentrated on studying patients with injuries to a limited part of the cerebral cortex, the so-called ventromedial frontal cortex. The interesting thing about these people was that they had some very characteristic social difficulties. There was nothing at all wrong with their intelligence as it is measured in terms of IQ, and in every conceivable psychological test of personality and ability, they come out in the normal range. But in their daily lives, the "ventromedials" were completely hopeless. They couldn't keep their friends, they couldn't hold down jobs, and they couldn't cope with their finances. All in all, it was as if the patients always made bad choices.

The issue was how this could be the case, since their ability to reason was apparently intact. After several years of research and the development of a special test, which involved a card game in which subjects could win or lose money according to how they dealt with risk, it was demonstrated that their rationality was missing a crucial element. Emotion. Intellectually, these brain-damaged people could easily understand that there was a risk involved in certain actions, but they didn't feel it "in their gut" like the rest of us and therefore did not react to it.

Damasio concluded from the series of studies that certain areas in our frontal lobes serve to connect emotion, choice and social behavior. And he developed what is called the somatic marker hypothesis, which says that our brains supply every choice we face with a sort of emotional tag that is reflected in physical emotion. Quite literally, we are talking about a "gut feeling."

The theory is still debated but, whereas neuroscience had treated emotion as irrelevant to the study of how we reason and make choices for nearly a century, Damasio and his colleagues have placed the emotions at center stage, hand in hand with rationality.

Here in Los Angeles, you sense that the new Brain and Creativity Institute is in its early days. When secretary Susan Lynch comes out to greet me, she has a dossier that I may look at but not take home with me. It is only sent out to people who might be interested in donating larger sums to research. I learn that the institute has close ties to the university's humanities faculty and that there are plans to forge a collaboration between neuroscientists and everybody from sociologists to filmmakers and pianists. Then, Damasio himself comes out to interrupt my reading.

He extends his hands in greeting and I am surprised by how small he is: a slender man with fine, bird-like features. With his quick, controlled movements, he also seems like a bundle of tightly-compressed power and energy. He is elegant right down to his fingertips. On this quite ordinary Tuesday afternoon, he is dressed in a neat, grayish-blue suit that can only be Armani and whose surface has a discreet sheen. He is also wearing a red tie and a cool, blue shirt. Not a thing out of place. Being with this elegantly-dressed man makes me wonder whether my skirt is on straight and whether he can tell that the polish on my toe nails is several days old.

"I feel frazzled," says Damasio immediately after an introductory hello. "Do you know what that means?" I reply that I know the feeling and suggest he takes a good, long vacation.

"Oh, God, I need one, but I can't. There's too much going on. When I was young, I imagined that there would be plenty of time to lean back and just think, when I got old. But then I got to head a large department at the University of Iowa and now I'm starting up something new here. We've got thirty people already, and the place will undoubtedly grow even more."

It undoubtedly will. The two Damasios are among the absolute neuroscience elite, the members of which jet around the world as invited guest speakers at meetings and conferences. "A coup for USC," Nobel prize winner and memory expert Eric Kandel called their move, underscoring the comment by calling Antonio Damasio "One of the great modern thinkers on the brain."

"I almost never give interviews anymore," says Damasio, declaring it to be a waste of time. "Journalists simplify the research until it becomes unrecognizable and make everything into a story about here's the brain center for this and the brain center for that. They want to sell an easy message that these 'centers' control and define us. The idea is easy to communicate, and it is easy for the public to swallow it, because it robs the individual of any responsibility. I can't stand it."

No, it's terrible, I pipe in, and the busy man smiles, satisfied, pulls me into his office, and sets me down on a sofa. On the table in front of me is a highly-polished bronze brain, whose medulla oblongata protrudes like a stalk from below. It looks like a dented balloon that could deflate at any moment, but according to the base it is a Golden Brain Prize. Just one of the countless prizes Damasio has garnered over the years.

"The term creativity must be understood in a special way at the center here," he says after sitting down in an easy chair that is quite a bit higher than my sofa. From these heights, Damasio relates in a well-considered and polished stream of words that the idea for the new center grew naturally out of his past work on emotion and behavior. He and his wife are, of course, quite interested in studying the neurology behind the creative processes in art and music, but it is first and foremost about social behavior.

"My definition is very broad," he admits. "The first thing about which human beings were creative was neither art nor technology but the creation of social relationships. This is the core of creativity for me. Our work has to do with how we create the patterns of human relationships and how we create patterns of social conventions and ethics."

The light in the office goes out suddenly, leaving us in twilight. It's because we are sitting still, Damasio explains, so the motion sensors in the room think it's empty. He lifts his arms a couple of times like wings, and the lights come back on.

"We have long been convinced that the next phase of our work must be to project neuroscience into social space. As far as I can see, the last decade's research points in a clear direction. We started by mapping the simple, elementary emotions – fear, anger, disgust – but quickly got into how emotion affects social behavior and social interactions. Now we have reached the point where the challenge is the decidedly social emotions."

Here, too, there aren't as yet many results but there is plenty of ambition. Damasio will be presiding over researchers who are going to dissect some of the complex feelings people have described from time to time as uniquely human. Advanced stuff such as pride, respect, admiration, generosity, contempt, shame, and guilt. Strangely mixed feelings that are difficult to define but play a huge role in shaping our existence.

"We want to know how these sorts of emotions are evoked in social situations, what happens in the brain, when we have them and what sort of bodily, physiological reactions accompany them."

This is easy enough to assert as a manifesto but more difficult to put into practice. Antonio Damasio doesn't go into details but explains that there will be a lot of studies of normal research subjects and of people with various specific brain lesions; injuries that affect certain emotions and, therefore, may indicate which brain cells these emotions draw upon. Another option is to put people in test situations and stimulate them to have certain emotions and then use scanning equipment to test hypotheses about which areas and networks of the brain are active.

"And what will this sort of understanding give us on a social plane?" Damasio asks himself. He answers after contemplating a moment.

"I think it's obvious that any knowledge about how the brain and the mind work in social situations will provide tools to deal with these situations. This is the general benefit. So, my overwhelming motivation is to understand more about human nature, and if you do that, you have the opportunity to be more reflective about your own conduct and to modify it. This is the general benefit of social neuroscience. Some would call it a poor benefit, but that is not true."

He stands up, inclines his head slightly to the side, and continues.

"If you look at the way diverse societies have dealt with problems like war and social conflict, particularly ethnic conflict, it is incredibly primitive. People have certain reactions that are emotionally driven and very poorly controlled and end up in irrational positions. And many of the problems we have dealing with these things have to do with people not recognizing the mechanism behind their feelings."

"They take them at face value?"

"Exactly. And they won't give in, won't lose face. That is what we're seeing now in Europe and its reactions to Islam."

"And just as much Islam's reactions to Europe?"

"Of course. Your country knows all about this," says Damasio, smiling pointedly. He refrains from mentioning the Danish cartoon crisis directly.

"Wouldn't it be nice if people in that sort of situation knew something about what they could expect from their own biology? And if they were capable of stopping and controlling their own reactions without a lot of ego and without imagining they're being humiliated?"

I can only agree. But, at the same time, I have to ask whether Damasio thinks that people might see it as provocative for neurologists to start making claims about ethnic conflicts. After all, they only study brains.

"One of the problems is that people believe that we neuroscientists 'only' study brains," he replies a bit impatiently. "But in cognitive neuroscience, we study the connection between brain processes and particular types of behavior. It's not the isolated lump of tissue that interests us. No, we are delving into the mechanisms of the conduct we see in the social world."

Damasio believes that neuroscience finds itself at nothing less than a decisive turning point. A point where it needs to come out of isolation and join forces with the social sciences and political theory. These are disciplines that have been good at describing social and political problems but have not been able to do much about them because of a lack of knowledge about fundamental causes and processes. On the other hand, when we understand more about conflicts, for example, we can draw a lesson that can be used politically.

And, personally, Damasio has a dream of helping the insight that is bubbling up from the field by channeling it back to the education system. "If I can stay alive," he says, looking like he'll be going strong for quite a while yet, "I have a goal of helping educate schoolteachers in neuroscience. Initiatives in education are just as important as initiatives in medicine. To this point, neurology and later neuroscience have contributed to the progress in the medical field, and I'm quite proud of the part I've played. But we are facing an important shift. In the future, the major consumer of our results will be society."

It all sounds very seductive but makes me wonder whether it is realistic. Will "society" – the broad masses – be capable of or interested in relating to the finer nuances of the brain's inner life?

"Oh, you make me think of a philosophical tradition that has, unfortunately, been lost. Philosophy was once a way of commenting on everyday life. Aristotle walked among the citizens of Athens and commented on how life should be lived and what you could do as a human being in the world. Even a figure like Spinoza, who was vilified and banned by the authorities, was actually known in his time among ordinary Dutch people. They talked about his ideas of freedom and about the separation of church and state and about the relationship between emotion and rationality. These were ideas that could be distilled and made accessible to ordinary people. Important facts and important ideas *want* in some way to percolate out of their creators and reach ordinary people."

But what are the important ideas? What are the intellectual products that are supposed to come out of the great factory of brain research? Damasio sits for a moment without saying anything but then provides his own distillate.

"Do you know what? If there is anything that can be said to be a central message with respect to the whole of biology, it's this: there is nothing especially moral about human nature as it has been passed along to us by evolutionary history. In nature as such, there is no good and evil. So, we can't trust ourselves and our urges and emotions as indicators for fashioning a society we call moral. We have to determine and define how we *want* to fashion society and our lives, regardless of what our nature dictates. It is up to us to build the structures we want."

Damasio holds up two impeccably tidy fingers.

"There are two tools to build with. On one hand, there is culture – accepted norms – and, on the other, our knowledge of our own biological limitations and tendencies."

Society in Our Heads

Man, oh, man. It's only a little after lunchtime, and I'm already existentially exhausted. I don't even have the energy to work myself up over an unfair ticket some overzealous representative of the campus police has placed on the windshield of my rented car while I conversed with Damasio. Instead of complaining, I report sluggishly to the office and pay up. I'm simply wiped out.

It's not because of the morning's nerve-wracking scanning and the subsequent relief that I'm not carrying around a brain tumor or a large inexplicable cavity. Nor is it spending the day driving around the crowded California freeways. No, it's the visions that these gentlemanly brain researchers have evoked. The smiling Iacoboni, who wants the neurorevolution to create better people. And the thoughtful Damasio, who wants to fashion social institutions from a deeper knowledge of the limitations of human nature. Decidedly ambitious causes.

But this idea that biology ought to play a role in politics – doesn't it remind you of something we've seen before?

Sitting behind the wheel, I vividly imagine how that sort of issue stirs up ominous feelings in one commentator after another. How they will say "remember eugenics" and wag their index fingers in the air. The comparison to the theory of race hygiene that had popular appeal from the late 1800s up to the Second World War is obvious. This produced social policies – and politicians – inspired by scientists who claimed a very precise knowledge of human nature. Unfortunately, the knowledge was mistaken. It claimed that human nature varied in accordance with race or population group. On the rationale that they were biologically inferior and would corrupt the quality of the population, Eastern European and Jewish immigrants were kept out of the promised land of America, and blacks and whites were kept separate in the South. Using similar arguments, people forcibly sterilized the mentally handicapped in countries such as Denmark and Sweden early in the twentieth century. And a bit later, the eugenic movement achieved its pinnacle in the systematic liquidation of Jews and Gypsies by the Nazis.

The conclusion follows of itself – keep biology far away from politics. But that is a trap we have to avoid. In reality, the call from Iacoboni, Damasio and the other brain researchers with ambitions beyond the lab is a much-needed wake-up call. It's not about scientists being cautious and keeping quiet but about the rest of us – society, opinion makers, politicians – staying informed and educated about advances in the field. We simply cannot allow ourselves to be indifferent to what is going on in brain research. It *is* a revolution, it *is* going on all around us, and it *will* have consequences for society in ways that we can now only imagine. If we want to help shape the influence this research will have, we have to familiarize ourselves with its language and be able to interpret its results; it is our responsibility to monitor its progress, asking difficult questions and providing intelligent criticism.

On the other hand, it is crucial in what we call a knowledge society that those who actually have knowledge enter the marketplace and test it. And we can already see knowledge about the brain entering into the debate. A prime example is the current US discussion surrounding single-sex education in public schools. Citing a growing body of evidence that girls' and boys' brains develop and mature at different rates and in some cases process information differently, advocates argue that gender-divided class rooms are a way to optimize learning for both sexes. The theory suggests that quieter girls blossom in classes without noisy boys and benefit from teaching that focuses on social connections and interactions. And noisy boys who are in danger of underachieving learn better when the teaching is designed to accept and even incorporate physical activity.

Scientists keep fueling the discussion with novel findings, as in 2008 when Doug Burman of Northwestern University was cited all over the media for an fMRI study showing that girls and boys aged nine to fifteen process language information using different brain regions.[32] The boys' performance depended on activity in visual and auditory areas while the girls had much higher activity in areas directly associated with language and abstract thinking. This was surprising and may help explain long-standing observations that girls are generally ahead of boys in language skills and women often score higher in verbal tests than men. But Burman didn't stop there. "Our findings could have major implications for teaching children and even provide support for advocates of single-sex classrooms," he said in a press release, explaining that boys may benefit from having language tasks taught both visually and orally while girls do equally well with either method.

Planning education on the basis of brain scans does not go down well with everyone. The high profile ACLU (American Civil Liberties Union) strongly opposes gender-divided schooling, calling it discrimination. More interestingly, there are proponents of gender division who specifically object to using the neuroscience rationale. This camp worries that explicitly telling children that their brains are so different they need to be taught apart would add to the prejudice and distance between the sexes.

So far, the science, coupled with practical experiences from schools over the last few years, seems to have made an impression on educators, parents and the powers that be. In 2006 the US Department of Education passed new regulations making it easier to create single-sex programs or even schools and today around 400 public schools offer the single-sex option. That is up from just a dozen in 2002.

The final word has not been said. There are lots of questions that could be asked of brain researchers, social psychologists and gender researchers. But *having* the debate is crucial. It is a discussion whose outcome will help determine what generations of young people will get out of their schooling. At the same time, it raises a general question: can you teach responsibly without taking into consideration how the brain functions? And as an extension of this: can you really fashion the best possible society by *not* taking into consideration knowledge about human nature?

6

ECONOMICS – THE INVISIBLE HAND OF THE MIND

If our innermost nature with all its limitations and idiosyncrasies is allowed free rein anywhere in life, it is in economics. And we're not just talking about mortgage equity, pension funds and cold, hard cash, but economics in its broadest sense, understood as the trade-offs, barters and financial decisions we make every day. Economics is anything having to do with making a choice, assessing alternatives, or planning a strategy – which, practically speaking, embraces every aspect of life. And it is in our endless haggling that we truly show who we are.

Just think of your private life: it is difficult to find an area in which you can avoid it. Parents are in non-stop negotiation with their children – not only about bedtimes and allowances but in a general exchange of love and care for good behavior and carrying on the aspirations of previous generations. Spouses run a brisk business, exchanging sex for tenderness. In the workplace, bosses invest goodwill and accommodation in their employees in the expectation that they, in turn, will reap rewards in the form of increased work efforts and a better bottom line. On the individual level, when we are – so to speak – on our own, we constantly weigh options, process information, and calculate possible futures.

Thus, economics is an excellent field to be in, if you are interested in human nature. And since economics deals with what people do in real-life situations, it is obvious that economists must work with the best possible picture of human nature in order to create workable models.

"Economics is a branch of biology broadly interpreted," wrote the well-known economist Alfred Marshall in 1890, and if that distinguished gentleman could rise from the dead, he would be pleasantly surprised to see that, after a century's delay, his thesis is being confirmed. Economics is slowly moving in the right direction. With neuroeconomics, it is about to become a part of neurobiology and the great neurorevolution. In fact, it is in the economic sphere that the knowledge of human nature provided by neuroscience has the greatest immediate potential – in the form of financial products and political measures – for changing society as we know it.

Neuroeconomics makes use of the approaches and techniques of brain research. Neuroeconomists represent a small, interdisciplinary advance guard that puts people in scanners and spies on their neural circuits as they play games and make economic choices and evaluations. Some are examining patients with brain damage and others are exploring the brains of monkeys. The intent, as one of the pioneers, George Loewenstein of Carnegie Mellon University, has put it, is "[t]o open the black box of the brain in order to identify the structures that underlie economic decisions."

And sometimes you get the impression that it's a Pandora's box that's been opened. At any rate, it must be said that the creature crawling out from under the lid doesn't resemble the easily-understandable, streamlined version of humankind provided by the classical economists. We are dealing with a creature who disregards logic, acts with tremendous irrationality, and has a contradictory nature. A character who loves gambling yet is conservative and risk-averse. Someone who seeks trust and thrives in it, yet enjoys punishing others when they offend his sense of justice.

This is something quite different from the traditional *Homo economicus*, with whom standard modern economists work in their models. This economic individual has quite a limited personality but is equipped with two reliable and key characteristics – absolute selfishness and complete rationality. He may not be scintillating or inspiring company, but his actions are entirely predictable.

Homo economicus was conceived in the early 1900s and grew out of a critique of classical economics, as it was practiced in the century before. At that time, economics was not considered a true or strict science but a sort of topic of discussion for philosophically-minded gentlemen. People interested in economics put out their ideas and theories to be debated in learned fora. However, the ideas were never measured against reality; there was no methodology for processing data.

Yet, even though they were not weighed down by research and statistics, the classical economists had a sense that it was important for them to take into account the nooks and crannies of the human mind with its myriad motivations. So, many of them grappled with sophisticated psychological concepts long before the discipline of psychology made its debut in history.

This was particularly true of Adam Smith, who today is often dubbed the "father of economics." He laid down the fundamental doctrines with which every economist works. However, when economists cite him today, they usually quote from his famous 1776 treatise *The Wealth of Nations*, in which Smith presents the concept of the invisible hand that guides markets and in which he introduces self-interest as the driving force of our transactions in the marketplace. There is much about rationality and selfishness in this book, but taken on its own it does not account for Smith's view of human nature.

As mentioned earlier, Adam Smith was a moral philosopher who maintained that we as individuals are guided in the choices we make in life by our "passions" – emotions, gut feelings, instincts. In *The Theory of Moral Sentiments* from 1759, he wrote about how human beings, by virtue of their innate moral nature, are concerned about each other. We are by nature equipped with altruistic traits that are manifested in the way we act. Deep down, we want to distribute worldly goods in a just manner and to take care of the poor and unfortunate. In other words, we are moral creatures guided by inner compassion.

These ideas seemed much too foggy and romantic a century later, when people like Marshall and others pioneered an economic theory based on the influence of behaviorism. According to this theory, people are not to be understood on the basis of ideas about compassion but judged by their conduct. Behaviorist economists cast a sidelong glance at biology. They believed that a thing's use value – *utility* in economic terms – must be reflected in processes that take place in the brain but that they could never be measured directly. The best we could hope for would be to deduce what was going on internally by looking at what was happening externally – that is, through actions.

Nice try, but still not good enough for later economists. They made use of another, even harder science as their model – Newton's physics. They wanted to be able to measure and weigh, and they ultimately wanted to present the world with a formal economic theory replete with mathematics and equations. Economics was to become a respectable science with rules and laws and simple, transparent models to explain the world. A breakthrough occurred in 1937, when Paul Samuelson published his treatise *Foundations of Economics*, in which he presented his idea of how to develop a simple, clear theory of human behavior. After Samuelson, mathematically-inclined economists were no longer interested in individual choices but in what aggregates, large groups of individuals, did – a national economy or a business, for example. And in order to understand something so complex, it was easier to have a simple model for the individual.

Homo economicus was born and he was doted on by his parents. After the Second World War, there was an almost explosive growth in this formal economic theory. More and more complicated theoretical models were built to explain and predict economic developments. The models presumed more and more rationality. Not only are all people rational, it was now asserted, but they all know everyone else is rational and they know that the others know it, too.

On the whole, *Homo economicus* was a success. In most general macroeconomic contexts, his rational and selfish behavior could be used to explain any number of things: unemployment, inflation, major national and international connections. But there were anomalies, things that didn't fit – in short, foolish behavior. Later, new empirical methods were developed and, with the computer and its possibilities for analyzing data, experimental economics arose. For the first time, you could test economic theories, and the reality check made for some harsh surprises in the ivory towers. As Harvard psychologist Steven Pinker put it:

"Economists find again and again that people spend their money like drunken sailors."

For some, a nagging question remained: why?

In 1979, Israeli psychologists Daniel Kahneman and Amos Tversky came on the scene. With all that rationality as a governing principle, the economic anomalies didn't make sense, said the two experts on the human mind. Kahneman and Tversky were inspired by perception research, which tries to learn about our sensory systems by studying how we perceive illusions and tricks. Their primary idea was to understand principles in cognition, decision-making and evaluation by looking at how people can be deceived with respect to the statistical principles by which, according to the theories, they should be guided.

Not long after the Israeli duo's challenge to economic theory, economist Richard Thaler of the University of Chicago went on a sabbatical to Stanford in California. There he met Tversky and became captivated by the psychological approach. The new comrades agreed on a vision for guiding the field of economics to a more realistic model – one that would take psychology into consideration but still be formal and have predictive power. The result was behavioral economics, which as the name implies takes its starting point in the actual behavior of people.

The field really took off in 1982 with an article in *Nature*, in which three German researchers introduced the so-called "ultimatum game" – a tool that was later used in different iterations again and again to examine human strategies in direct economic exchanges. In the game, two individuals face each other. A researcher comes in and gives one of them a sum of money – say, ten dollars. That person, the "proposer," must now offer the other player, the "responder," part of the money – at least one dollar. The responder must choose between agreeing to the bargain and walking away with the money or rejecting it, which results in the researcher taking back the ten dollars – in which case both players leave empty-handed.

It is not difficult to predict how *Homo economicus* would handle this situation. In the proposer's role, he would, as the selfish pig he is, offer the smallest amount – one dollar. If he were in the responder's position, he would – as a thorough-going rationalist – accept the offer, acknowledging that even a miserable bargain is better than none. However, real, flesh and blood people react differently. They have an idea that it is unjust for the proposer with a bulging wallet to offer so little, and they reject the low offer. For his part, the proposer has the same sense of propriety and insight into the responder's feelings, which makes him offer something more "reasonable."

Ultimatum has been played in various cultures all over the world. The typical outcome is that the proposer offers around four dollars and the responder reacts with "yeah, well, okay." There are societies in which significant deviations from the norm may be seen – for example, certain tribes in New Guinea – but they can be counted on one hand. In fact, there is only one group that deviates consistently and markedly from the norm, and this group behaves exactly the way *Homo economicus* would have: people with autism, a group characterized by having no insight into or understanding of other people's feelings.

As for the rest of us, economists had no immediate explanation for the source of our irrational behavior. Evolutionary biologists, however, came to their aid by claiming that this odd strategy might derive from primitive hunter/gatherer societies in the past, when such behavior was in and of itself fairly rational. If you rejected small monetary gains, your standing rose, and you got a reputation for not being an easy mark. An advantage that could very well pay off in the struggle for survival.

Now, stories from the dark recesses of evolution are always entertaining, and many are even plausible, but a need arose to verify what was going on in living, contemporary people, biologically speaking. Behavior is all well and good, but a small group of researchers wanted to dig a little deeper – people like the American economist Colin Camerer and his colleague George Loewenstein. In 1997, they put together a two-day inspirational conference in Pittsburgh at which neuroscientists, psychologists and a handful of economists were to inspire each other to new sorts of investigations. This was the germination of neuroeconomics. It was the natural successor of behavioral economics and, with neuroeconomics, people began digging around in the heads of the economic players.

For example, in 2003, eleven years after the first ultimatum game, a team of psychologists led by Jonathan Cohen of Princeton had their test subjects play the game in an MRI scanner.[33] Thus, they could register what a person feels when receiving an "unjust" offer. Throughout the game, there was brain activity in the dorsolateral prefrontal cortex, which indicates cognitive information processing and planning for the future. However, as soon as the proposer made an offer that was viewed as unacceptably low, blood began to stream through a completely different area – namely, the anterior insula of the cerebral cortex, an area whose activity is connected to strong negative emotions such as pain and disgust.

It seemed to the researchers as if the two areas of the brain were in a struggle for control, as the research subject was considering her reaction. At any rate, it turned out that those who rejected the offer had greater activity in the insula than in the prefrontal cortex, while the reverse occurred in those who chose to accept the smaller sum.

The result is strikingly reminiscent of the observations of Joshua Greene – and his mentor Jonathan Cohen – on moral choice. Emotions fight with rational impulses on decisions, and it appears that the emotions are decisive. This led Colin Camerer and George Loewenstein, who are two of today's superstars of neuroeconomics, to reformulate Plato's old metaphor. The Greek philosopher described the human mind as a charioteer, whose chariot is drawn by two horses – reason and emotion. This is true enough, say Camerer and Loewenstein, with this important difference: reason is a pony and emotion is an elephant. In fact, one can find the notion of an internal tug of war in Adam Smith. He speaks of our conduct as a struggle between the "passions" and the "impartial spectator."

In fact, these two can be seen battling it out whenever we have to decide between an immediate gain and delayed gratification. Plain old run-of-the-mill psychological studies revealed long ago that people generally prefer an advantage here and now, even though it would be more profitable – and thus more rational – to wait. And in yet another scanning study, Cohen and Loewenstein demonstrated that there are separate neural systems involved in these two types of decision.[34] Test subjects were given a number of choices with respect to a sum of money. For example, they were offered five dollars here and now or forty dollars to be paid six weeks later. The researchers toyed with greater or lesser sums and longer or shorter delays to see how the parameters affected the decision and the brain activity.

In general, the prospect of realizing a particular advantage at once created a great deal of activity in areas of the limbic system that deal with emotional stimuli and such things as expectations of reward. So, here-and-now appealed powerfully to the emotions. On the other hand, emotions were difficult to engage in situations when a person had to wait a long time for the money. And as soon as there were two choices out in the future, the constant activity in the cognitive areas in the prefrontal cortex suddenly weighed in the heaviest. Logic won out and the person chose the larger but more distant reward.

You would expect that the reverse would also be true – that while we prefer to get a positive reward right here and now, we would want to put off something negative for as long as possible. Classical economic theory also predicts this. But in fact, the reverse is often the case. If people are facing a choice between getting a negative experience over with quickly or putting it off until later, many choose the former.

A team of psychiatric researchers from Atlanta's Emory University recently wondered how this could be explained. Their hypothesis was that economists failed to take into account the fear and trembling factor, and this parameter must be incorporated as an independent part of the overall equation. With the young doctor and researcher Gregory Berns in the lead, the team gave terror a form in a fascinating experiment in which test subjects were given electric shocks.[35]

The thirty-two volunteer participants were put in a scanner, where they received a small test shock on their feet so that they knew what to expect. In the first part of the real experiment, they were told when they would receive a shock and how bad it would be. In the next part, they could choose between a powerful shock delivered immediately and a somewhat weaker shock after a thirty-second delay. So, which would it be?

Twenty-three chose a lesser delayed shock, but there were nine – almost a third – who without a moment's hesitation preferred a strong jolt at once. Berns and the others saw an incredibly interesting difference in the brain activity of the two groups. In the group which preferred the immediate, stronger jolt, there was a conspicuous increase in activity in the areas of the brain responsible for the perception of physical pain. However, these areas, which go under the collective name of the cortical pain matrix, showed no increased activity in the other participants.

You can see many of the neuroeconomic studies as an indication that rationality and logical calculation are the thin layer of varnish coating a series of automatic reactions. The affective contribution, which has to do with motivation, desire and disinclination, is something we *feel* strongly but to which we do not have conscious access. The sense that something is just supposed to be a certain way is a result of hidden calculations in the depths of the brain. You could say that "our" conscious calculations are packed on top of impulses from the unconscious and must compete and argue with them. In a rather beautiful way, this system reflects a developmental history.

Our emotions have been handed down to us through a long evolution in which they have served as quick, effective decision-making mechanisms. Automatic reactions that, from a statistical point of view, worked well and were fine-tuned over time. Conscious cognition is a much later invention, and it was not until the advent of our incredibly developed prefrontal cortex that a massive leap in capacity was achieved. Yet, despite all its capacity, this new mechanism did not take over the decision-making process. It is this surprising teaching that has come out of neuro-research over the past decade – especially from work done with people suffering from brain damage, whose cognition and intelligence are intact but who are unable to engage emotionally. Without feelings, our choices fall apart and actually become irrational. The passions are a necessary engine in this intricate machinery.

One of the trump cards of the emotional brain is the chemical reward it can provide by activating the dopamine system. We all know the satisfaction that can come from monetary gains – a stock that rises, a premium bond that comes due, or a winning hand in a late-night game of poker. Such feelings are all the result of a dopamine jackpot. Interestingly, you may score a bit of the same when you choose to cooperate in a social context. A number of studies from the infancy of neuroeconomics have looked at the so-called trust game in which two players are given a sum of money that can be multiplied more or less according to the degree the players trust each other to share rewards instead of keeping it for themselves. The game has been played both person to person and person to computer, and when it is played person to person, researchers typically see activity in areas of the striatum in the basal ganglia. These are regions packed with dopamine neurons, whose stimulation is associated with enjoyment.

The same reward system kicks in when it comes to our darker side. You can see this in one of the most interesting contributions of neuroeconomics, which reveals that we enjoy punishing cheaters. A rational economic person would only punish others if she herself received some material benefit from it, either directly or as a deterrent. Reality shows that we gladly punish, even though it may cost us in purely material terms. Some modern economists call this "altruistic punishment" and, of course, it can be seen in action directly in our heads.

In a now classic experiment, Ernst Fehr of Zurich University used brain scanning on a group of research subjects who were playing a money game.[36] They sat across from each other, two by two – A and B – beginning with the same amount of money. A could decide to give B a sum of money and, if B returned the money, both players would be rewarded with an extra payment. However, if B decided to keep A's money, A could punish B. The punishment might be purely symbolic, costing nothing – a simple verbal wagging of the finger – or both players could lose everything. The researchers found that a failure on B's part always made A deliver a punishment – not symbolic but real – even though it meant that A also lost money.

The reason for this could be gleaned when the players were examined with MRI scanning. The images revealed that the punishment decision unleashed significant activity in brain areas that are involved in enjoyment and satisfaction. It is the dorsal striatum that typically becomes active, when we experience a reward after goal-oriented behavior and effort. It was also the case that the higher the striatum activity the individual research subjects showed, the more money they were willing to lose to punish the cheater.

A more recent related experiment indicates that there is a difference in how much men and women enjoy wreaking vengeance upon those who have cheated them. Tania Singer of University College in London had research subjects play the same game Fehr had played – the so-called prisoners' dilemma – with a series of partners. Some of the partners cheated flagrantly and reaped the reward, while others preferred to cooperate. Afterwards, the research subjects were scanned as they watched a video in which their various partners were apparently punished with electric shocks. And the disturbing scenes led to widely different reactions.

In each case, the researchers could see an increased activity in regions that have to do with the experience of pain and which, at the same time, have mirror neurons that react at the sight of others' pain. However, identification or empathy with the victims was quite dependent on how they had behaved in the previous game. The sight of the nice partners being punished caused much more activity than seeing the cheaters punished. The most striking thing, however, was that the male players not only had less empathy with cheaters, but they apparently enjoyed seeing them suffer. Unlike women, they also showed increased activity in the enjoyment areas in the striatum and the orbitofrontal cortex of the cerebral cortex.

There is no agreement on the extent to which the experiment reflects a biological rather than a cultural difference. It may be that the women in question were raised and socialized to be more gentle creatures, while their male counterparts were allowed to give vent to the whole register of human emotions. However, these observations undeniably shed new light on how we maintain norms in a society. It may be that we as individuals do not directly gain anything by making those around us stick to the rules, it just feels good to punish norm breakers.

On the other hand, like the tendency to adapt oneself and follow collective rules, trust is a social lubricant, if not a fundamental ingredient for society to be able to function at all. And trust is one of the phenomena that neuroeconomics has pounced upon in recent years.

Among the areas of focus is the hormone oxytocin, which has an effect on the brain and has long been known to be a substance that strengthens social bonds. In nursing mothers, oxytocin flows freely, just as orgasm releases a shot of the drug in both men and women. People have theorized in this connection that oxytocin serves to bind the two sexual partners together in order to increase the chances of survival for their offspring. But not only that. Oxytocin appears to be able to make us trust strangers so much that we are prepared to share our money with them.

Ernst Fehr and his colleagues in Zurich have seen indications of this in their research. They had students play an investment game with each other.[37] One – the investor – is given a certain amount of money of which she can decide to give a portion to her counterpart – the trustee – who immediately has the amount tripled. The trustee can then decide how much he wants to give back to the investor. The more the investor trusts the trustee, the greater the chance he has to get back his original investment. But the more trust, the higher the risk that the investor will lose, if the trustee decides to cheat her.

Fehr had half the young investors take six nasal inhalations of Syntocinon after playing a few rounds. The nasal spray, which contained oxytocin, had the effect of causing them to increase their investments significantly in relation to the other players who were not exposed to the hormone. Through a number of controls, the researchers showed that this trust did not have to do with a general desire to take more risk. No, those affected by oxytocin were specifically willing to accept the kind of risk that comes from social contexts. One could call it a hormone with a social slant.

A single chemical ingredient is one thing, but what is trust in a social context? How is it expressed, when you go beyond the conduct that is there for all to see and delve inside the brain? It appears to be identical to an expectation of reward. This is what a team of American researchers concluded from a particularly ambitious experiment in which they connected two MRI scanners in Pasadena and Houston. From their respective corners of the continent, Read Montague of the Baylor College of Medicine and his colleagues Steven Quartz and Colin Camerer of the California Institute of Technology put their volunteer subjects into what they dubbed a hyperscan.[38]

Through a computer connection, two players were put in contact, while their brains were observed by researchers online. In the scanners, the research subjects played investor and trustee; but instead of just playing one round, they were asked to invest and get returns and relate to each other's reactions over ten rounds. On the built-in screens in their goggles, they could track both how things were going with their investment and how the other person behaved. The investor could adjust her confidence in the trustee in accordance with the way he had administered the investment in the previous rounds, and the trustee could read how much the other person trusted him and adjust his transactions accordingly.

Forty-eight hyper-couples participated in the study, and the researchers could subdivide their exchanges into three types – neutral, positive, and negative – in which the parties respectively rewarded and punished each other's transactions. Generally speaking, they came down hard immediately if the other party let them down and, in the next round, sent less money through the ether. In the same way, trust was rewarded promptly with more money.

In the corresponding scanning, one area of the brain was prominent – namely, the caudate nucleus. This little kernel, the size of a peanut, is found deep within the striatum and is known to go into action when we expect some form of reward. Numerous studies have shown that it lights up when people are promised a glass of juice or a bag of money. It is also activated when cocaine users take a hit and when gamblers win at roulette.

Throughout the rounds of the game, the activity of the investor's caudate nucleus could be seen to fluctuate according to how much she decided to invest in the trustee. In other words, the activity was a measure of her trust. The same thing could be seen in the trustee, whose caudate nucleus activity also fluctuated up and down according to how much he was given by the investor and how much he then decided to give back. You could even observe the two players form an opinion about how trustworthy their counterpart was. Gradually, as the game progressed, researchers could see the trust signal of the two brains appear increasingly quickly. Toward the end, the activity appeared fourteen seconds earlier than in the introductory rounds.

The spectacular hyperscan received a tremendous amount of press. For one thing, people were a little taken aback that it was the first time anyone had ever observed the intimate details of a direct social exchange. At the same time, the researchers' news was also a little surprising. We usually think of trust as something utterly non-utilitarian, something you give without shady ulterior motives, which is its own reward. Sadly, research is now indicating that what we call trust apparently reflects nothing more than the expectation of being rewarded.

Neuroeconomists and the Avuncular State

"Yes, the hyperscan experiment was really cool," says Colin Camerer, when I meet with him on CalTech's almost tropical Pasadena campus. "And by the way, we found some interesting gender differences. Men made their trust 'calculation,' sent off the money, and then stopped thinking about it. The brain activity in the areas involved went like this ..." He makes a diving hand gesture. "On the other hand, for some time after the decision, women continued to have increased activity in the areas that express expectation of reward and areas regulating concern."

"Are you saying that women think too much?"

"I wouldn't put it that way. But maybe they think about the consequences of their choices and actions a long time after they've made them. Longer than men, at any rate, or ...." Camerer looks as if he would rather talk about something else.

"But," he says quickly, "right now I'm more interested in curiosity. We are just about to submit our first paper investigating the phenomenon. It'll be going to *Science*," says Camerer, sounding suddenly like he's talking to himself. "It might go straight through, because people think it's interesting and new, but then again the referees might say it's pure bullshit that doesn't have anything to do with anything."

He doesn't seem to take it all that seriously. Colin Camerer is in his mid-forties, heavy-set with a beer belly stretching a blue T-shirt to near breaking point and a wreath of tousled grey hair surrounding a balding peak. I think of Andy Capp but know that

Camerer is sharp as a razor. Unusually gifted, he was accepted at Johns Hopkins University at fourteen and acquired a reputation as something of a mathematical prodigy. In his free time, he made use of his abilities at the race track in his hometown of Baltimore, and moving in gambling circles stimulated his interest in probability calculations and choice. This then led the young man in the direction of economics and, at twenty-one, he graduated from the prestigious University of Chicago Business School with a Ph.D. Later, he became known as one of the driving forces and original thinkers of neuroeconomics.

"Curiosity," Camerer says once more. "I think of it as a hunger for information."

We have gone to his spacious university office but remain standing at each end of it. I wonder how long it will take before Camerer thinks of offering a seat.

"Our experiment involves asking people a series of questions – such as what country in the world has the most women in government."

"Sweden," I say nonchalantly. I have the home field advantage here.

"Wrong! It's Finland."

Camerer is one of those people who are always completely focused on whatever they're talking about right this minute. His eyes are intense and his eyebrows have a habit of constantly moving up and down, as if they were participating in the discussion and trying to draw attention to something.

"The research subjects are scanned, as they get the questions, and we see a characteristic pattern of activity. It takes place in cognitive areas in the dorsal frontal cortex, which as expected indicates that they are thinking and processing information. But something is also going on in the caudate nucleus, which as you probably know ..."

"Expectation of reward?"

"Exactly. And this is what makes me talk about a hunger for information. Anyway, we believe the person's degree of activity in those areas reflects how curious they are, and this is what's interesting. We can see that people who are very curious but answered our questions wrong reacted with significant activity in the areas surrounding the hippocampus."

"Something to do with memory?"

"Precisely. Apparently, they start encoding information into memory. And in later tests, they were far better at remembering the answers than the others. We believe that hunger for information actually increases our ability to remember and learn. Understood in this way, people learn more from their mistakes and store new information more effectively, if curiosity is stimulated than if you aren't quite so curious."

The hypothesis is certainly interesting, but how it relates to economics seems a bit difficult to fathom.

"I'm thinking along the lines of internal motivation among knowledge workers. Classic economic theory always talks about work as if someone is lifting crates, but we live in a knowledge economy. So now it's about organization theory and knowledge workers. Studies like this will have significance for education and training in general. It has to do with acknowledging the fact that people have to be curious in order to learn optimally and finding ways to stimulate this curiosity."

Finally, Camerer realizes that we are still standing up and begins to look around and to clear a space for us to sit down. There are piles of papers on every flat surface in the office, but he clears the seat of a small chair and pushes it to the corner of the desk for me. I sit down and discover a bright orange rubber brain right in front of my right eye. It looks almost like a football and I squeeze the spongy mass a couple of times, while my host turns his back to remove something from his own chair. When he sits down, he takes the conversation in a completely different direction.

"I see neuroeconomics as a natural extension of behavioral economics," says Camerer, explaining that it goes into the questions in more depth. Instead of speaking in general psychological terms, the researcher asks directly where in the brain these things take place in an actual situation. I play Devil's advocate and ask whether you really get something extra out of observing the brain that you don't get just studying behavior.

"That is the $64,000 question, and there are many ways to answer it. On the one hand, you can argue that, no matter what, someone – ordinary neuroscientists – will be interested in economic subjects ..."

"Exactly!" I mention a quote by Camerer's old mentor, Richard Thaler. The prominent behavioral economist said to *Business Week* that he didn't believe neurostudies added anything in particular to behavioral economics and that economists should leave neuroscience to the neuroscientists. Camerer leans forward in his chair.

"We absolutely *cannot* leave it to the neuro-people. They don't have a proper understanding of game theory, and they have no idea what games are good models for which real life situations." His eyebrows are dancing, and he shakes his head. "The neuroscientists also believe that the core of an economic decision is money. But the central thing is trade-offs. You only have so much time and so much money, and you always prioritize things, since everything has value. Money is just one of them."

He leans back and asks rhetorically, "But do we need to understand the brain processes in relation to economics?"

And the answer – in economics jargon – is that the best argument for neuroeconomics is its option value.

"It's just hard to believe that something usable won't come from looking at economic questions with neurotechnology. I compare it to going to the Moon. For centuries, people were incredibly interested in the Moon and they tried to get knowledge about it through telescopes and analyses without ever thinking to go up there – it just wasn't possible. Then somebody suddenly said, let's go up there. Skeptics could say that, strictly speaking, you didn't need to, but think of the return on investment that came out of it. Technologically and knowledge-wise."

Camerer opens his arms wide: "Of course, we have to study the brain in economics!"

And for those who don't feel convinced, he points out that there are still tons of mysteries in every important area of economics. No one knows why stock prices and stock markets fluctuate the way they do. And with respect to saving, no one has any idea why there is such an enormous difference between countries.

"Here in this country, people don't save anything, nothing, while the residents of Singapore hoard up a fourth of their income. Lots of economic models say that twenty-five percent is ideal for the national economy, so why don't we save anything here in the US? Is it because we are bombarded with credit opportunities? I mean they stuff tons of easy credit offers in the mail every week!" His eyebrows become one unmoving line. "What's the savings rate in Denmark?"

"Uh, let me see ..." I don't know the numbers and reveal my ignorance of economics.

"Well, you guys are probably doing fine. But here – there are a million personal bankruptcies in the US every year, and it may well have to do with these credit opportunities. It might also have something to do with the values that are instilled in us from when we are kids. Maybe, something to do with the fact that American kids are bombarded with commercials from the time they are infants and don't develop impulse control the way people do in other places."

"Aren't there also individual differences in how frugal we are?"

"Of course, and where do they come from? The ultimate thing would be to study these sorts of questions by looking at behavior *and* doing neurological studies. That could reveal what the brain perceives as reward and the motivation in different individuals and different cultures."

I ask Camerer whether he believes all this will change our view of human nature.

"Absolutely." He is silent for a brief moment and continues with a somewhat slower diction. "But isn't it rather that we are finally getting to know what human nature *is*? The combination of genetics and brain research will provide a lot of insight – and especially knowledge about nuances and variation – that we don't have today. Just think of the fundamental questions about the extent to which we humans are 'good' but victims of temptation, or whether we are fundamentally 'bad' and therefore need to be limited and controlled. Who is who? And how can we identify the difference? They will be able to uncover much more about this sort of thing in the decades to come."

Camerer suddenly ducks down and disappears under his desk. He returns with a liter Coke bottle from which he takes a massive swig.

"And then there is what might be the most important point of all. I think we'll discover that the nature of the individual human being is far more mutable and malleable than we think today. There is lots of evidence of plasticity, especially in children and young people, and I'm convinced that we will develop strategies to increase plasticity later in life. Do you know Richard Davidson and his monk studies?"

"Of course." Now I feel on more solid ground again.

"These monks significantly change their brain waves and way of thinking through training in mental techniques. And I'm sure you can boil these old techniques down to far fewer, more effectively targeted training sessions and still have an effect on a number of aspects of the personality and one's perception of the world. In particular, there are the possibilities for making mental changes that will make people happier, more satisfied."

It is strange to hear the word happy come out of the mouth of an economist. Traditionally, this lot have concentrated on recommending strategies and actions for the health of general systems rather than human satisfaction. But a new tone has entered into the profession. In recent years, people have been talking about a new paternalism, a benevolent guardianship, piggybacking on the understanding of our psychological architecture provided by behavioral economists and neuroeconomists.

Among today's new paternalists are the American economists Samuel Bowles and Herbert Gintis. These two former Marxists have renounced their earlier position and done various notable studies in behavioral economics. Today, for example, they vociferously advocate changing the welfare system so that it "does not clash with human nature." Among other things, this implies that systems give up automatically supporting just anybody who happens to be in need. Those who receive assistance should *deserve* it. And this is so, because those footing the bill need to know that the recipients are not the cause of their own problems and that there is an intention to improve their situations. If this is not the case, according to Bowles and Gintis, you may conceivably wind up with something that resembles the need of the strong to visit altruistic punishment upon the weak.

"I don't need to teach you about paternalism. You know all about it in Scandinavia," asserts Colin Camerer. "But elsewhere it has a bad reputation. Economists have a strong innate reluctance to do anything that conflicts with individual self-determination unless that determination causes harm to somebody."

You can almost hear the echo from John Stuart Mill's essay *On Liberty*. It asserts that the state should only intervene in the actions of an individual if they are harmful to others.

"No doubt, economists are uncomfortable with the concept of paternalism, but as I said recently to a journalist from *The Economist*, the danger of harmful paternalism does not come from behavioral economists but from religion. Just look at this country! The compulsion to regulate human behavior is greatest in the religious right, and they are the ones screaming the loudest about freedom and all the terrible consequences of the nanny state."

In its modern incarnation, the nanny state is being promoted as a libertarian paternalism that still wants to preserve free choice. The goal of the new paternalists is not to create an overprotective

Big Brother, but what has been called the Avuncular State. It is the vision of a state apparatus that, with knowledge of the individual's best interests, provides a gentle nudge in the right direction – just to make sure that we don't succumb to our own lamentable foibles.

"I see it as a reasonable challenge to behavioral and neuroeconomics," says Camerer. "If we can see why people make bad or irrational choices, we should also be able to define some guidelines for how people can do things better."

We are living in times of great behavioral challenges – one example is the much talked about global obesity epidemic that is threatening to overwhelm health care systems. People have been bombarded with information that they must eat less and exercise more. Everybody knows this, and most would even like to do something about it, but in practice it comes to nothing. It is about willpower and self-control. And it is this sort of phenomenon that behavioral and neuroeconomists can begin to study and which their models, as opposed to the classical models, can speak about. So the question is whether the combination of real challenges and the new scientific possibilities will inevitably lead to more regulative policies.

"I'm convinced of it," says Camerer. "At any rate, we've had a far more thoughtful discussion in recent years, and that's good. Honestly, paternalism will exist no matter what you say or do. There are rules for when you can start drinking, when you can start having sex, when you can vote, and when you can join the military. And where do these rules come from?"

He knits his hyper-mobile eyebrows close together over the bridge of his nose. "What I'm talking about now won't happen tomorrow and *shouldn't* happen all at once. But over time, I can imagine, for example, that the conditions for joining the military will be regulated by something other than an arbitrary age limit. Today, you have to be eighteen; then you can enlist and maybe end up getting killed on the battlefield. Maybe there should be an evaluation of your ability to make decisions and assess circumstances before you make that kind of decision."

"Do you mean mental tests beyond the normal intelligence test?"

"Today you have a lot of admission requirements that are purely physical – how healthy you are, whether you're in good shape – but maybe it would be better to have a test of specific mental competencies. You can imagine a test of judgment. Both to protect young people who can't evaluate what they're getting into and to avoid problematical personalities with psychological or neurological defects that could have a negative impact."

At this point, I would like a more concrete example. Camerer looks up at the ceiling and thinks for a moment.

"You could imagine that, in a few decades, potential recruits might get a brain scan while they're shown pictures of civilian dead and of families grieving over a dead relative. And if their brains don't react to the stimuli in the right way, they don't get in."

In other words, a peek into the coils of the brain will keep immature boys and psychopaths away from legal weapons. Undoubtedly something many people might be pleased about, but the methods will make some people nervous.

"I know it scares people. To a lot of people, putting somebody into a scanner sounds eerily mechanical, reminiscent of mind-reading. But if you think it through, it's no scarier than the present system in which it's simply about being eighteen years old!"

Colin Camerer himself doesn't seem nervous about the prospect of taking action on the basis of the knowledge brain scanners will be able to produce in time. He draws a parallel to health research.

"There is a lot of interest, for example, in doing something about drug addiction, which has to do with the brain's reward systems. As economists, we are also interested in investigating whether there is behavior we could call economic pathologies. Today, we're dealing with things and we don't really know whether they're an expression of something outright abnormal."

He mentions shopaholics and workaholics.

"That people are workaholics who really can't let go of their work may seem fine on the surface. They can get things done. But being a workaholic is actually a really bad thing in terms of stress – it certainly leads to illness and family problems, right? And it can ultimately result in huge costs in the health sector."

If, in the future, workaholics were to be treated for and protected against their unhealthy tendencies, Camerer believes, it might also be a good idea to protect society against other potentially harmful deviations.

"You can imagine that we in research will find particular effects in the brain that might relate to judgment or particular prejudices or something else entirely that might have less than optimal effects on a person's decisions. It is possible that these effects are small and insignificant, if the person has some sort of average job. But they might be extremely harmful if they were allowed to unfold in key positions in complex economies – UN Secretary General, President of the United States, chairman of the board of Enron, whatever. If you could test for special brain types in certain positions, it might not be so bad."

We both agree that we are probably a long way off anyone proposing the testing of the neural capacity of an American president, but there are other important posts where that sort of thing might not be so inconceivable.

"Wall Street," says Camerer, smiling broadly. "I've just met with a bunch of finance types, and they are very interested in the brain."

Then there are those who aren't so interested. The President of New Frontier Advisors, Richard Michaud, who administers hundreds of millions of dollars, believes that neurofinance has no place whatsoever on Wall Street. And Alan "Ace" Greenberg, who was a legendary and longtime CEO of the investment bank Bear Sterns has gone out of his way to say, "All this neuro stuff is just horse shit, when it comes down to brass tacks."

"Okay, okay, the field is divided," says Camerer. "There are those who get the idea and those who don't. But generally people who work in financial markets are far more interested in psychology and brain research than academic professors. Finance people *see* inexplicable and remarkable things happen every day, while professors sit staring blindly at tired old theories."

But while they do that, stockbrokers and fund managers juggle billions of dollars. Wall Street, investment firms and banks all over the world are rolling dice with other people's money. And as Camerer says, "It is quite important that they do their jobs optimally."

"Think, for example, of the hedge funds and investment firms, where individuals make enormous trades and wagers. The firms worry about what we call *rogue traders*, people who spin out of control and engage in wild transactions," says Camerer, suddenly holding the Coke bottle tight against his chest. "There was recently a case about a man who all by himself lost six billion dollars in a single fund. Him, we could have done without."

The question is whether it will be possible to scan to find out who is dependable and who is a Chicken Little.

"You can imagine it happening at some point. It's interesting – on one hand, you want people who are to some degree immune to the pain of financial loss, but at the same time they can't be so immune that they take too many risks. But gradually as we get a better handle on what neural mechanisms determine the balance, we can, of course, do something about identifying them."

Another obvious topic for study is people who really understand how to make money on the world markets: Wall Street wizards, stockbrokers who notoriously earn huge sums for their clients and private speculators with phenomenal fortunes. Everyone knows the story of George Soros, who, with the right amount of risk tolerance, speculated against the British pound and beat even the Bank of England, walking away from a single day's trading with a billion dollars in his pocket. According to his son Robert, Soros changes his position in the market according to certain spasms in his back. But you can imagine wanting to look into whether something special is also going on in the head of this kind of super-investor.

"Absolutely," says Camerer. "Here you've got some people making decisions based on the same information as everybody else but who are different in some decisive way. They do something other people don't do in the same situation, and now we have the tools to begin looking systematically at how their psychology and the processes in their brains actually distinguish them."

Investment Brokers with Lizard Brains

One person who has earned quite a bit of money over the years is the professional investor Warren Buffett. Presently the richest man in the world, he has become known as the Oracle of Omaha and long ago achieved the kind of legendary status that makes him a ubiquitous commentator on the world's financial media. According to one of his most well-known oracular statements, it's no problem for an investor with an IQ of 25 to be successful. What it takes isn't superior intelligence but rather, "A temperament that can control the urges that get other people in trouble in investing."

There are indications the man is right. A small circle of researchers has begun studying good and bad investors, and the first studies are now in. It began in 2002, when Andrew Lo, an economist at MIT, was given access to one of Boston's largest investment firms. He chose ten currency traders at various levels of seniority and wired them up with electrodes to take a measure of their emotional weather. Lo recorded the traders' heart rate, breathing, blood pressure, body temperature and even how much they sweat during the day's trading, playing with hundreds of millions of dollars.[39] Just a normal day at the office. But as the electrodes revealed, it was a day filled with emotional explosions. Hearts were pounding, palms were sweating, and there was hyperventilation as soon as there was volatility in the market and exchange rates fluctuated wildly. In particular, very quick movements tended to tickle the traders' sensitive nervous systems. But one thing that jumped out at Lo was that the more experience the individual had, the better he was at keeping his own emotional roller coaster under control.

At first, the professor wasn't allowed to check the actual success rate of the traders and could say nothing about who was actually the best performer. For that he had to wait another three years. However, when he had a group of traders with somewhat uniform experience come to the laboratory to invest under controlled conditions and demonstrate their abilities in simulations, it turned out that the most successful were those who were effective in keeping their emotions on a somewhat even keel – both when they were earning money and when accounts were hemorrhaging. The fewer the emotional swings, the greater the total earnings.

Andrew Lo isn't just a professor. He is also the director of the investment firm AlphaSimplex Group. So he has more than just an academic interest in getting as far as possible in his analyses of advantages and mistakes. And as he recently told an interviewer from Bloomberg Markets, he would like "To delve deeper into the brain to understand what is going on and to refine our methods in the real world". So at MIT, he has started scanning selected expert investors to see whether their brain activities reveal special characteristics.

The people who at the moment have delved deepest into the investing brain are Brian Knutson and Camelia Kuhnen at Stanford University. There were reverberations when, in the fall of 2006, they published an article in the prestigious journal *Neuron*[40] in which they could point to what types of brain activity predicted good and bad decisions in ordinary amateur investors. Kuhnen and Knutson had nineteen volunteers, both men and women, play investor with twenty dollars in their pocket and the possibility of investing it over several rounds. In each round, they could choose between a boring but rock-solid bond that would yield one dollar each round and two stocks, each with its own risk profile. There was the "safe" stock that had a fifty percent chance of yielding a profit of ten dollars and a twenty-five percent chance of a loss of ten dollars. In the risky stock, on the other hand, there was a fifty percent chance of losing ten dollars and only a twenty-five percent chance of gaining the same amount.

In the scanner, the nineteen subjects weren't told which was which and had to rely on the actual performance of the two stocks throughout the experiment. That is, they had to learn from experience which were the riskiest choices and which paid the best dividends. But as the two researchers coolly observed at the beginning of their article, these average investors systematically deviated from rationality when they made financial choices. In the specific experiment, they could see two types of deviation from the optimal strategy – which is to say, the strategy that one would expect from a rational player with a healthy, neutral relationship towards risk. There was a risk-seeking cohort and a risk-averse group and, twenty-five percent of the time throughout the game's investment rounds, the investors made mistakes that were characteristic of their individual strategies.

Prior to each mistake, the researchers could see an equally characteristic activity. Each time a risky investment was made, and particularly when a risk-taking choice was made, there was extra activity in the nucleus accumbens, where euphoria resides and one is rewarded with dopamine. You could say that the expectation of reward here is expressed as greediness, and it made the excited investor ignore the inner calculator. On the other hand, when the investors made a safer choice or when they froze up and made risk-averse mistakes, the anterior insula lit up. This harbinger of pain, disgust and negative emotions in general seemed to make them recoil with fear of failure and loss.

Outside of the laboratories, the financial markets hail the principle that logic is king when it comes to investment. When people deal with securities, they make their choices based on expectations created by a competent, well-considered analysis of the information available. The theory of rational expectations even resulted in a Nobel Prize. It went to Robert Lucas in 1995 and he is still at the University of Chicago, where he teaches that people decide with their intellect and learn from their mistakes.

"This is pure nonsense. If the neuroeconomic studies show anything, it is that we human beings are full of innate prejudices and oddities that make us poor investors," says Richard Peterson. He is a psychiatrist, and he has granted me a telephone interview at seven o'clock in the morning. He runs Market Psychology Consulting in San Francisco, whose home page explains that "[D]ue to the fault of evolution, we are not wired optimally for making money in the financial markets."

Peterson, who is still under thirty-five, has experienced this first hand. When they were in college, he and a friend ran a small investment firm that traded in futures and did financial projections. To his own astonishment, young Peterson discovered that "emotions and the unconscious always trip up the right decisions." When he was making projections, he was right in seven out of ten cases, but as soon as real money belonging to real people was involved and he was making actual trades, his success rate plummeted to a measly thirty percent.

"Because I couldn't figure out what was going wrong, I couldn't really do anything about it. But that is exactly what I think the field of neurofinance will allow us to do," Peterson says today. He has two kinds of customers. There are fund managers who are rolling in success and would like to know what they are doing right, so they can train their colleagues to perform equally well. And there are financial workers with problems; stockbrokers who have run into choppy waters and are incurring great losses or who suddenly find they can't handle risk.

"I only prescribe medicine to such clients if there are clinical symptoms, not to achieve optimization," stresses Peterson, who doles out anti-depressants now and then. "But in fact, the normal thing is to give them an introduction to what we know about the brain and the way it makes decisions. I explain studies such as Knutson's and show them the scanning images, and the insight into the mechanisms helps quite a lot."

This is a sort of cognitive self-therapy – based on the principle that simply knowing your amygdala is playing tricks on you can weaken its power.

"But it only works to a certain degree. The mechanisms it deals with act on an unconscious level, so there needs to be a sort of alarm system for the individual to react to when it comes into play."

Peterson imagines a future in which stressed-out finance people are not only attached to their cell phones and computer screens but also to smart, portable EEG-measuring systems. Systems that will be able tell when you are moving into a danger zone, where investment prejudices or unfortunate psychological tendencies might backfire. And which could be used to systematically improve results.

The financial sector is interested in this sort of vision, Peterson feels, but he also notes quite a bit of skepticism. Market players want to see some actual products, something they can buy which will give them a measurable return on investment.

"We could offer to scan potential analysts and traders in order to assess whether they have problems with risk or too little self-control but, as it stands today, it would provide no more information than what you could derive from a psychological test. The scientific studies still only scratch the surface, and we need to understand more about which neuronal circuits are involved in what types of situations, and how they vary among individuals. In the same way, we need to know much more about when the various ways of making decisions are practical and impractical. It may be that a particular mechanism will lead to losses in a rising market but can function very well in a declining market."

But when this sort of knowledge is available and can be tested with a visit to an MRI apparatus, Richard Peterson has not a moment's doubt that it will be used as an entry ticket to a job on the stock market. And when will this happen?

"I'd say, give it ten years, and there will certainly be a product."

Does this seem frightening? A little shocking, perhaps? Worries always seem to surface when intriguing new knowledge begins to take shape in the form of concrete technology. Suddenly it exists in the real world, and it doesn't sound altogether comforting to think of a future in which you can get access to a person's innermost being and reveal intimate parts of their personality just to decide whether they're right for a job.

But is this new? Not at all. Beauty is only skin-deep, as the saying goes. We have always wanted to go deeper. To be able to read the psyche of other people, to get beyond their defenses, to evaluate their potential and predict their reactions has been vitally important in an evolutionary context, and today it's worth gold on the employment market, a market in which we are already testing each other. In virtually every business, there are psychological checklists and personality tests galore that are administered to future and present employees to ensure that the company only hires those optimally suited to the job. No one can seriously imagine employing or promoting key employees simply on the basis of a good CV, a pleasant face, and assurances that the job is just right for them. No, you have to know who they really *are*.

The products young Peterson imagines with such radiant vision will undoubtedly find their way to the market, because the test mentality is already well established and because current methods are lousy. The interminable check-off forms with their sometimes bizarre questions and obvious cross-references might well provide a hint about general personality type, but whether they provide a deeper understanding than an ordinary conversation backed up with a competent knowledge of human nature might yield is doubtful. With brain research's ever-deeper understanding and the scanner's ever-clearer picture of the processes going on within the skull, you get a powerful up-to-date tool to tackle an old task, but a tool that in time will be able to investigate and test practically every possible facet of our mental abilities, characteristics and states. Today, it's the ability to make decisions, tomorrow it may be loyalty, and the day after susceptibility to bribes. In fact, a team of European and Australian researchers showed in the spring of 2007 how an MRI scan can reveal with ninety percent certainty whether a person suffers from OCD; obsessive-compulsive disorder. That is just the beginning. The simple existence of such a tool opens up considerations of whether it should be used in cases such as those Colin Camerer so briskly brings to mind: a critical examination of heads of state, the World Bank, or military personnel with responsibility for vital operations or enemy prisoners. Is this frightening? Perhaps. But is it more frightening than closing our eyes and hoping for the best?

The battle lines for discussion have been drawn, and we will soon be debating where the boundaries are. How deeply can you look into an individual's personality, and who is to have access to the information? Should insurance companies see it? What minefields and uncertainties lurk in the head of consumers? Could parents investigate the potential of their children before they decide to spend a fortune on an elite education? Might future spouses check on their sweetheart's proclivity for fidelity before taking vows?

These are interesting questions. And beneath them will undoubtedly run a wider discussion about the neuroeconomists' new brand of paternalism. I'm sure Colin Camerer is correct that this idea of a benevolent, helpful Avuncular State has a bright future. It follows naturally when you begin to look at human beings as clever machinery or systems that are subject to certain tendencies and a particular repertoire of behavior. Inevitably, the urge will arise to make use of this knowledge to shape and improve things. When you know the fools out there are incapable of saving sufficiently for their old age and will be coming to you with their hat in their hand, shouldn't you prepare for this in advance?

Of course, seen from the outpost of the British or other European welfare states with their lifelong automatic payments to the retirement system and obligatory employer pension programs, this sort of nanny-ism isn't shocking news. But in more freedom-worshipping parts of the world like the US, it is new. A slightly exotic idea. And yet – we are already seeing this mentality take root in ever more drastic non-smoking policies all over the world and in the campaigns to make us eat more vegetables and fish and to protect ourselves from the harmful rays of the sun.

Soon the neuroeconomists will be able to contribute knowledge about how to exploit the brain's innate tendencies to give us that extra push from the inside. From there, it is only a small step to politicians "helping" us do as we're told – for our own good, of course, as well as for the greater good. And there we are – suddenly, Antonio Damasio's vision of linking neuroscience with politics has been realized.

7

SELLING IT TO YOUR NEURONS

Say the word: *neuromarketing*. Doesn't exactly sound good, does it? It's an outlandish word that scrapes across the tongue, leaving an aftertaste of thought control, science fiction, and downright creepiness. The press surrounding neuromarketing reflects this as well. The headlines are ominous: soon, the bright boys of the advertising world will get their sticky hands on our inner "buy button." Soon, marketing experts, with the help of cutting-edge brain research, will get direct access to the inner depths of our brains where, with the right stimulation, they can unleash our buying impulses and get their cash registers ringing.

Neuromarketing is a young and growing field – some won't even admit that it *is* a field yet – that is striving to reveal the inner mechanisms of our consumer behavior. You might say that this interest and the issues it raises are a natural extension or offshoot of neuroeconomics and the more general studies of how we make choices and decisions. Every so often, there is also a conspicuous overlap between neuroeconomists and researchers in neuromarketing. The studies in neuromarketing are just more specific and much more directed. And the Holy Grail lies in predicting *what* the brain wants.

In the advertising industry, you can see neuromarketing as an attempt to make the "art" of advertising into a science Any marketing expert proposing a multi-million dollar project to a client would like to be able to back it up with something that looks like real data, not just hunches. To answer this need, marketing has already drawn on psychology in developing tests and theories, and ad people have borrowed the idea of the focus group from social scientists. Brain research is the third wave. And neuromarketing has taken on a warm, fuzzy glow in the advertising world, where they convene meetings and conferences about its potential and, every so often, proclaim in their journals that it is the undeniable wave of the future. Such enthusiasm is harder to find in the scientific arena. Marketing is *not* a science, many say, pointing out that only a small handful of studies have been published in scientific journals.

Still, the whole thing started in academic circles, when in 2003 Clinton Kilts of Atlanta's Emory University called in a team of volunteers for a series of experiments to throw light on the brain's role in product preferences. How does activity in brain cells mirror things we are crazy about as opposed to things we absolutely hate or that just don't speak to us? At that point, Kilts had nothing to do with marketing or advertising in general, but the fundamental question tickled his fancy.

The volunteers came in and, in the first round, were presented with an array of various consumer goods, which they were asked to rank by appeal. Simple answers on a numerical scale. In the next phase, they were taken through the MRI scanner as they were once again shown the same goods, while the apparatus registered the brain activity they aroused. When Kilts later analyzed the reactions of the research subjects, there was a common feature that leapt to his notice at once. Every time one of them – male or female – saw a product they really liked, blood rushed to a little area towards the front of the brain. The medial prefrontal cortex lit up like a beacon in the images.

This result lit a fire under Clinton Kilts, who knew he was onto something interesting. The medial prefrontal cortex is not just any old brain region – it is an area very much involved in our self-identification and the construction of our personality in general. This part of the frontal lobes is involved when we relate to ourselves and to who we are in some way. Kilts was quick to draw his conclusion. The scanning experiments, he believed, indicated that, if you are attracted by a product, it is because you identify with it. That the product fits into the picture you have of yourself.

This was quite exciting – in a nice academic way – but the debut experiment seemed to provide an obvious opportunity to do a new sort of study of the market. Kilts could see a future where researchers didn't have to go out and ask people what they thought about a product anymore, or rely on their vague answers and poor self-insight. No, potential consumers could just be scanned and the answers could come straight from the brain.

Not long after his breakthrough, Clinton Kilts helped to found a new division for the American marketing consultancy BrightHouse, their Neurostrategies Group. Their focus was not intended to be ordinary market studies of the type that are supposed to tell producers how to put together a commercial for strawberry jam or sports cars to hit a target market. It was claimed in their launch statements that all the studies done would be of a general character – designed to increase our understanding of how consumers think and, in particular, how they develop a relationship to companies and brands.

The discussion quickly came to turn on the concept of branding. The fact that something – be it a product, an institution or a concept, for that matter – is not just immediately recognizable but has a narrative of its own. The product is not just a physical thing but comes with a whole mental universe that penetrates the consumer. Think of Gucci, iPod, Mercedes, and take note of the images the words bring to mind. Branding has been a hot topic for a long time in the advertising world, and it is one with phenomenal force. Most of us know that branding palpably influences our choices and shopping habits, but researchers suspect that branding can also fundamentally change the way we comprehend sense impressions.

At least that is the obvious conclusion to be drawn from the only (so far) classic study in neuromarketing, a fascinating study of what can be called the Pepsi paradox. For decades, it has been known that Pepsi is the preferred cola in blind taste tests, but it is still Coca-Cola that continues to be the absolute bestseller in the US and the rest of the world. However, since 2004, we have been able to see the short-circuit going on in the head of the cola-drinking masses.

The originator of the experiment was Read Montague of Houston's Baylor College of Medicine, who must be credited with breaking through to the broader public with the experiment, which was essentially a cola-tasting while being subjected to MRI. Just under seventy volunteers were first asked to taste the competing products in a blind tasting and, just as so often before, Pepsi was the big winner. Pepsi also set off greater activity in the so-called ventral putamen than Coca-Cola. The putamen is an area cradled deep in the brain in the striatum, which is, among other things, a component in the reward system. So, the interpretation was straightforward – the activity meant "this feels good."

In the next series of experiments the subjects tasted colas with visible labels. When the research subjects knew which brown liquid was which, almost all of them suddenly preferred Coca-Cola. They were convinced that the taste of Coca-Cola was far superior to Pepsi. This shift in attitude followed an important change in the brain – this time, the medial prefrontal cortex went into action. The cerebral cortex intervened with its higher cognitive processes and triumphed over the immediate feeling of reward that was evoked by the taste impression. The product that actually tasted worse and provided a poorer physiological reward was viewed as better when the whole identification apparatus and the idea "this is so me" went into action.

The cola experiment, which came out in the journal *Neuron*,[41] might be said to show that branding is mind over matter. And, of course, this got marketing people to think in a new way. Now they could hope that the methodology of brain research would help to explain how people build up the much sought-after positive branding story. The dream is that researchers with their scanners will discover what has to be done to get the right elements into play to achieve a tenable branding. Storytelling aimed right at the medial prefrontal cortex.

And a small handful of them are going at it full steam. It is reported that, in exchange for a quarter of a million dollars, BrightHouse will test how products are viewed by consumers and figure out which of the product's attributes encourage the most self-identification and thus the greatest inclination to buy. On its homepage, BrightHouse boasts of having well-known brand names such as Coca-Cola, K-Mart and Delta Airlines in its stable. But you don't get any closer than the homepage. The people at BrightHouse don't answer the phone and are known for keeping their cards close to their chests. Other firms, such as the Austrian

ShopConsult and Neurosense in Oxford also tout neuromarketing, and the latter will gladly talk shop.

Things are going so well, according to co-founder Gemma Calvert, who has a background in both marketing and neuroscience, that they don't even need to advertise. Neurosense simply cannot keep up with the demand coming from a broad range of industries – based, she believes, on word of mouth. The price in Oxford is also quite different from in the US, costing between £35,000 and £55,000 to do a study using fMRI – not much more than you would shell out for a corresponding traditional survey with focus groups. And these groups can be very wrong. Calvert experienced this with a client who wanted to test a new product for branding – "I can't reveal who it is, because they're a little sensitive." The scannings at Neurosense indicated that the public was not enthusiastic. Nonetheless, the product was launched in the US, where focus groups had given it the thumbs up – and it was a total fiasco.

For Gemma Calvert, the projects are typically about finding an objective physiological measure for our reactions to products. She is in the middle of a major collaboration with one of the big names in the perfume industry, trying to figure out what scents really do to us. The producers or the "noses" who develop scents like to throw around terms like "soothing," "exciting," or "refreshing," without really knowing where they come from. In a sort of clinical test of a handful of scents and ingredients, Neurosense is looking at how they affect the brain activity and physiological parameters of the research subjects. You *can*, according to Calvert, see that some ingredients, for example, seem to be soothing and to mute activity, while others increase attentiveness, and the goal is to be able to label scents with a declaration.

While the perfume industry is new to their client stable, the media industry is a familiar face – including major players such as Viacom, which owns quite a few TV channels, and PHD Media, which plans and purchases commercials for clients from all industries. When it comes to TV commercials, they have always been bought and paid for solely on the basis of how many viewers there were at a given point in time, but this turns out to be a pretty poor method. With scanner studies, Calvert has shown that the context in which a commercial is shown means everything for how it works. For example, by measuring the response in emotional regions of the brain, you can see how a particular context can be positive or negative in the overall reception. For example, showing a campaign for the Red Cross in the middle of an episode of South Park is downright damaging for the way the individual brain in front of the screen views this charity.

But when it comes to campaigns, Neurosense is trying to make a breakthrough in the public sector. Great Britain has a number of messages it wants to impress upon on its citizens. They should avoid drinking and driving at Christmas for example, and of course, they should stop smoking. But how to drive the point home? Here, Gemma Calvert's MRI scanner can be used to scan smokers to pinpoint which campaign films and warnings on cigarette packets are most off-putting. There seems to be interest. At any rate, Calvert has arranged meetings with "several public agencies."

In the public sphere, you also find enterprising university researchers who use their academic platform to develop neuromarketing concepts. One example is Steven Quartz of the California Institute of Technology. Along with all his academic studies of economic and moral choices, Quartz developed a marketable package solution for the film industry which he noted is struggling with a measly four percent profit margin, and decided they could use some help. What is first and foremost on offer is precision – both from the films themselves and from the trailers that sell the films. Whereas you normally show these to a test audience to observe their reaction, Quartz offers to put the same audience into his scanner.

The idea is that the scanned feedback will provide information that the individuals cannot voluntarily provide, because they simply do not have access to it. In particular, Quartz looks for what he calls memorability. It is crucial for a trailer, which you might see months before the film hits the theaters, to leave an impression that is not erased as soon as the film stops running. And while you can't ask a test audience how well they will remember a trailer, a scanner can provide a clue to this by revealing whether the film initiates activity around the hippocampus, in brain areas that have a crucial significance for storing new impressions in long-term memory.

Similarly, Quartz would like to refine his methods to the point where they can say something about what is characteristic about a given stimulus and what the brain, therefore, takes special notice of. A great deal of our unconscious brain activity is spent trying to filter impressions from our surroundings and deciding what is sufficiently important to be brought to consciousness. But interrogating people about this sort of thing isn't very effective. For example, how do you answer a question such as: "Tell me everything that was interesting about the trailer to *Rocky IV*?" But greater knowledge about what kind of activity patterns determine which details slip through the mental filtering system could lead to the development of a trailer according to what is most likely to work its way into the viewer's consciousness and be remembered.

Of course, it doesn't have to be limited to movie trailers. Not unexpectedly, neuromarketing also has offshoots in the political arena, where it's all about selling your message and, especially, your candidate. In 2004, when the American presidential campaign had exploded into personal attacks and negative ads, some of the frontrunners decided to explore the depths of Republican and Democratic brains. Two former campaign strategists for Bill Clinton, Tom Freedman and William Knapp, allied themselves with Freedman's brother, psychiatrist Jonathan Freedman, and his colleague Marco Iacoboni from UCLA to find out how these two groups of voters reacted to the available election propaganda.

In the very same MRI scanner that tested my own mirror neurons, the researchers played a campaign ad to a group of volunteers and saw that Democratic brains are disturbed to a far higher degree than Republican by signals of danger and threats. The images of the attacks on September 11 on which the Bush campaign relied heavily produced much higher activity in the Democrats' amygdala – a sign that they found it far more negative than their Republican counterparts. The same difference was present when they showed the two political groups a video from 1964, in which the Democratic candidate, Lyndon Johnson, used the threat of nuclear war with its ominous mushroom cloud. For the armchair politicians Freedman and Knapp, the result confirmed an old theory that Democrats react more strongly to the use of power than Republicans.

But what's the use of this stuff? Yeah, well, that's still a bit foggy. However, when the media immediately pounced on their attention-grabbing experiment, Freedman and Knapp were careful to rule out using neuroscience to create a more manipulative campaign spot. No, no, not at all. They just wanted to focus on exploring this "new frontier," because they were really interested in what was going on in the heads of voters. In other words, they wanted to introduce a little more "science" into "political science."

Later, in 2006, they came out in the *New York Times* to say that they actually didn't believe politics was ready for neuroscience. Then, with the consultant firm FKF Applied Research, they began to offer their services for the study of business advertising methods. And business is apparently booming, because the two discovered that the vast majority of the countless commercials to which we are subjected every day make no impression at all. According to FKF, there is simply no measurable reaction from the brain to half the commercials on American TV. These expensive seconds are processed no differently than any other average sense impression.

On the other hand, there seem to be some clear winners that are able to engage viewers. In 2006, supported by FKF, Marco Iacoboni and his colleagues Jonas Kaplan and Eric Mooshagian performed what they called an "Instant-Science" experiment in which they analyzed how the commercials for the American Superbowl game did in competition with each other. Five research subjects were shown the short advertising spots, and the next day the researchers put the results on the Internet. The clear winner was Disney, with an ad encouraging people to go to Disneyland, that was able to create a positive expectation and empathy in otherwise unsuspecting viewers. At any rate, the advertisement set off a lot of activity in the reward areas of the striatum and in brain regions containing mirror neurons. Lagging behind Disney came an ad for Budweiser, followed by Burger King which only managed to excite activity in the amygdala, indicating fear and loathing.

Still, the question nags: how much science is there in such studies? Take the Superbowl experiment and the response to the Budweiser commercial for example: it's about a man drinking beer. The sight of the man spurs a response in the viewer's mirror neurons, which might indirectly indicate empathy or identification. Or it might just indicate that there are some mirror neurons that happen to react to seeing a certain type of movement. "But whatever it is," writes Iacoboni on his homepage, "it appears to be a good response to the ad."

The scientific community, however, is not impressed. Neuromarketing "looks to be a very speculative investment," as it was put in an editorial in the journal *Nature Neuroscience*. And Princeton neuroscientist Jonathan Cohen observes that, while brain scans offer incredible possibilities for looking into the brain, the interpretation of the results can be extremely complex. He believes it can be difficult to avoid attributing to them whatever it is you are already looking for.

And there are other killjoys on the field – consumer activists who, as a matter of principle, don't care for the methodology of medicine being linked to the advertising world's objective to increase sales. They were quick off the mark. Shortly after Clinton Kilts entered into an agreement with BrightHouse, the consumer group Commercial Alert of Oregon sent an open letter to the Chancellor of Emory University, asking him to sever connections with the marketing people and deny them access to university facilities. They had a hard time seeing how an institution that is seeking cures for diseases could defend lending its precious equipment to research into how to sell more cookies and cola to the masses.

The university thought everything was in order. The researchers over at BrightHouse were doing research that they intended to publish in scientific journals. And management couldn't see anything wrong in taking things a little further down the road from inquiry to invoice. As the dean of Emory's medical faculty said at the time to the *Atlanta Journal-Constitution*: "It's clear that, if you understand how people make decisions, there is a commercial application to it."

"And is there anything wrong with that?" asks Read Montague over the telephone from Houston. As the busy head of a large research center, he doesn't have a lot of time to give me but wants to emphasize that he wasn't part of unleashing an uncontrollable monster into the world.

"Yes, we sparked things off with the cola study," he says by way of introduction. "It showed that cultural messages *can* have a great effect on decision-making, just as it showed that you can test questions of product preferences directly on people and get an answer through scanning, without having to ask people directly. As far as the commercial aspect is concerned, you can say that this sort of method is obvious if you're BMW with a new car model to go out on the market. You might very well want to test what happens when you modify the wheel rims or the headlights in a particular way. Does it look like it might push potential buyers in the right or the wrong direction? If I were a businessman, I would think it would be a reasonable decision to test it with some scans before I risk millions and millions more on product development."

"You see neuromarketing as a way to avoid expensive mistakes?"

"Clearly. A lot of product fiascos are pumped onto the market and why not avoid them if you can? It's something else whether you can use the technology to get people to buy all kinds of crap. That's the 'buy button' that the media all talked about and trot out at regular intervals as a horror scenario. I believe the cause is Ralph Nader and his people at Commercial Alert, and I also believe it's a complete misunderstanding."

"Let me be clear about this," says Montague, making himself very clear. "We *have* no buy button! People aren't robots. We are the only animal in the universe who will die for an idea. People will go to war or stage a hunger strike for an ideology or a faith. There is no other animal capable of suspending basic biological urges for something they've intellectually decided to do."

"Neuromarketing isn't as scary as it's made out to be?"

"No. Research into neuromarketing, or decision-making and communication, which it actually is, will not turn us into mega-shopping zombies. On the other hand, it is likely to help us figure out *why* we shop, why we buy something particular, and how different groups of people can be influenced by the market, by cultural messages and the way they are delivered."

Then, Montague invokes a very American concept. What this particular field of research will lead to, he says, is *empowerment* – not helplessness.

"If we can do a scanning experiment that supports the conclusion that children have a tremendous reaction to finding small bits of toys in their burgers, then you can suddenly take a political position that, if children are to avoid getting obese and getting diabetes, you should maybe prohibit small bits of toys in burgers. I don't think it will lead to incredibly more effective ad campaigns but rather to a society that actively decides to set some rules for how advertising can be conducted."

"It is also voiced in academic circles that they can't really see what marketing research has to do with universities."

"Yes, but they have completely misunderstood it. This way, the results come out in the open, where they belong. I believe the worst that could happen would be to remove this type of research from an academic environment, because then it will happen in the private sector, where we have no say in whether new important knowledge comes out. Look, we're being armed to resist the influence others have on us for Christ's sake! Knowledge is always power, and I don't know how you feel, but the more we know about how we can be controlled by external influences, the more we can resist. And reflect."

"Why do I shop?"

"For example. But if people have an immediate aversion to this stuff and describe it in such scary terms, it's because we are entering into an area that is only now being explored – namely, the brain, the self and the mechanisms of it all. People feel in a way that it's like rummaging through their drawers for dirty underwear and they don't like it."

Montague draws a parallel to reproduction. Once there was a time when people knew very little about the details of where babies come from and, while superstition and bizarre ideas flourished in science, the subject was taboo in the ordinary discourse of people.

"But then they did the research and laid the cards out on the table and what was once furtive and naughty was now a completely normal part of the culture and the collective conversation, and today we create babies in the lab. This exhaustive knowledge of our own reproduction processes changed us. Very definitely. And it is just as certain that exhaustive knowledge about the brain and the nervous system will change our view of human beings and humanity."

"How?"

"Precisely how is difficult to predict. But I think it will be in the direction of more self-understanding and thereby more power over our daily lives."

We're just playing

Nor can Marco Iacoboni, the man who talked to me before about an existential neuroscience and human nature, see any problem. When I return to his corner office at UCLA to hear about his research in neuromarketing, he is not shy.

"Aw, come on," he says, smiling as broadly as before. "People buy things all the time. They love to look at products, and, on the whole, they enjoy commercials as part of the entertainment."

Then the slim Italian straightens up in his office chair and looks more serious.

"As I said to you before, I believe that neuroscience needs to get out into society. And the idea of using brain scanning to understand what people like appeals to me as a simple way of getting out of the ivory tower and into the market."

But the *science*, I say plaintively, is there any science to it? The area has produced so few publications they can be counted on one hand. Iacoboni cocks his head.

"The thing is that this sort of study doesn't fit well with scientific publications. For the most part, they are not designed to understand a biological problem, but to understand a relationship to certain products. But listen. We know that people have no idea about how their own decision-making process works or about how they actually make choices in life. They can tell you all kinds of stories about how they prefer one thing to another and why. But in reality there is no connection between their inner decision-making processes and what people *think* is going on."

Just like at my last visit, Iacoboni's hand finds its way up to his chest at the same time his voice becomes a bit darker. Almost solemn.

"We believe," he says, "that neuroscience and brain data can provide a much better prediction of human conduct than people's own explanation."

"But do you have any idea about the extent to which the brain activity you point to as a positive response is actually reflected in people going out and buying products?"

There is a short pause in the conversation, nearly imperceptible.

"No. We have no idea. I suppose you could test it by advertising differently in different states and looking at the result, but it would be tremendously difficult to do."

"But this means that the companies and ad agencies that are plopping large sums on the table don't really know what they're buying," I say, wondering whether that's going too far. Iacoboni just laughs out loud.

"They have no idea, but they're eating it up raw! But remember that all the traditional research into advertising is pure guesswork; it has nothing to do with science."

I continue to push my luck and ask whether neuromarketing isn't actually just something advertising agencies can market to big companies. Iacoboni nods vigorously and declares himself "in complete agreement." He even thinks there is an imitation effect out in the market.

"There is *hype* around scanning studies and, when the agencies see other people using them, suddenly they all want them, because they're afraid of being left behind."

Once again, Iacoboni becomes serious.

"No, we don't know whether specific conduct can be linked to what we see in the brain. Even in Montague's classic Pepsi experiment, he's looking at something subtle. Does he know whether it's the branding of Coke that's selling it? No, right? Listen, I'm deeply interested in the phenomenon of imitation, and we know that people imitate others and let their view of reality be influenced by others. In the virtual market, for example, we know – from a neuroeconomics study – that people who download music from the Internet do so to a high degree from the knowledge that *others* have also chosen that music. So, it may not be so much about the brand and the narrative and all that, but just about whether *other people* like the brand or the music."

Here, I think it's time for a quote I've memorized for the occasion. Neuroeconomist George Loewenstein says that neuromarketing is like reading tea leaves.

"Okay, I don't know if I can go along with that. We look at people's brain activity and, from everything else we know about the brain, we interpret the results and say what we think they mean."

For the UCLA group, this meaning is linked to what they have learned from their scientific studies of mirror neurons. These special cells are part of the whole package.

"We know that mirror neurons are crucial for empathy and we believe that these cells must in some way be behind our ability to identify with someone or something. So, if we see an activity in areas where they exist, it can help predict whether people identify with a product," explains Iacoboni.

That sounds harmless, but I have to ask, playing Devil's advocate, whether his mirror neurons really say anything at all. If there is ultimately an identification, it is only one part of the whole decision-making process. I can easily identify with the beautiful young girls who drive fast cars in the commercials, but that doesn't mean I'm going to get up from my recliner and run out to the nearest Mercedes dealer, credit card in hand.

"Of course not. That's why we're also looking into what's happening in the reward system and other areas that are markers of human conduct. But we have some pretty clear results in some of the experiments. Let me just show you something," says Iacoboni, who begins to tell me about an experiment he and young Kaplan are performing for one of the big credit card companies.

"I can tell you the name, but you can't write it down, okay?"

That's perfectly okay, I assure him magnanimously, whereupon he turns his Mac so I can see. In the experiment, Iacoboni shows the subjects two series of pictures in the scanner. In one series, the secret firm's name and logo are plastered all over building facades or on T-shirts and, in the other, you see the same scenes without the logo, but with the credit card itself placed strategically.

"See, there are some distinct blobs in the scanning. Now we're looking at the difference in activity in the two situations. First, you have the scenes with the logo and there is, of course, activity in the visual cortex – the red blobs in back – and almost nothing else."

It's easy to see – some dark grey brains, each with two blood-red blobs toward the back near the neck. I nod. Iacoboni smiles cheerfully and puts the other series of pictures up on the screen.

"When they see the scenes with the credit card itself, something else happens. A completely new activity appears in the front of the brain. The premotor regions. And this makes sense, because there are mirror neurons that become active, when they comprehend things that can be gripped and used. But it also means for me that the brain is responding much more by seeing this *thing* than by pure symbols. So, my very simple advice to the company is that they shouldn't just put their logo out in the public sphere but that they should display their product, the card. It's much more stimulating."

"Maybe, they should show someone using the card? Wouldn't that be even more simulating?"

"It could be," says Iacoboni, lost in thought about his pictures. "But this is a dramatic difference. Just look at the blobs!

"I'm working with neuromarketing, because it allows me to do some fun experiments that I otherwise would not be able to get funded and which can't be published in the scientific literature. But they can provide insight into how the brain works. Maybe we'll discover something really important and fascinating; nobody can know."

He says he has a dream that he hopes he can get some businesses to support. It has to do with finding out what activity in the brain acts as a marker for decision-making.

"I'm sure there is brain activity that, in reality, is better at predicting people's behavior than any statement they make themselves. I would *really* like to find this type of activity. It's a helluva job. I guarantee that there will be different markers that play a role in the decision-making process, all depending on what it has to do with. Whether it's buying a car, getting married, and so on. There isn't a centre for decisions. But the brain makes calculations by drawing on different things in different areas, so in some decisions there may be greater weight on emotions, while in others there is more weight on cognition."

Before he loses himself entirely in the future, I ask him about his outings into the field of political marketing. It turns out that Iacoboni, Kaplan and one of the Freedman brothers are coming out with an article based on just such a study.

"I'll print it out, even though the most interesting thing isn't in there. You know the fundamental design – it was in the middle of the 2004 campaign, which was very polarized, and we started just after Kerry's nomination. We got hold of the registered voters from both parties, people who had decided whom they were going to vote for. We first showed them the three candidates – the independent Ralph Nader was a sort of control – and then a Bush ad with powerful pictures from September 11. Then they got pictures of the candidates again."

We leaf through the article and note that the opponents' likenesses are connected with activity in familiar emotional areas that indicate distaste and negative feelings. There was also activity in some cognitive control regions, when the research subjects looked at pictures of the opposing candidate. However, this cognitive input is entirely absent when they view their own candidate.

"This can be interpreted that they feel something negative about their opponent and try to stifle it cognitively. *Or* it can be that they are using cognition to increase the distaste they already feel. They think about what an idiot their opponent is, or how stupid his policies are."

But what about that identification he's always talking about? What happened to that? I ask whether there is anything to see in the mirror neuron areas and sound, perhaps, a little sly.

"Hah. I had a feeling you'd ask about that. And yes, when the subjects viewed their own candidate there was actually a major response in the orbitofrontal cortex, which is associated with positive stimuli and empathy, and which is anatomically connected to mirror neurons. But listen to what happened. In the middle of it all, the *New York Times* came and interviewed us and did a front page story on the experiment, and all the hoo-ha had the result that we had to wait three months until everyone had forgotten about it, before we could go out and find the final participants."

And now the images looked different.

"What is fascinating about these social situations is that they change over time and are not like experiments that have crystallized in laboratories. The campaign had become really vicious and negative over the summer and, in our last group, we saw no empathetic response with their own candidate at all. No response from the mirror neurons! We did a formal analysis between these two groups, and there was an enormous difference. I think the result indicates that negative campaigns may work in the short run, because they repress the empathetic response you otherwise have for your own candidate. At the same time, I think the strategy can be incredibly dangerous in the long run, because it makes voters distance themselves from politicians as a whole. And is that something you want?"

Iacoboni has been speaking very passionately but checks himself before his enthusiasm runs away with him. Yes, yes, he admits these are interim hypotheses and extrapolations, but they are definitely interesting. And they kindle his desire to do more political studies.

"More than anything, I'd like to figure out what makes people go out and vote. The US has a huge and embarrassing problem that people don't vote. It would be an extremely difficult experiment to set up, and I haven't done it yet. But it will definitely not be a classic design in which you propose one fully-formulated hypothesis against another and test them. It will be a more open study that has to do with gaining insight into how the brain works. It will provide an imprecise picture, but the more we do of this sort of thing, the more precise it becomes."

The gentle whirring noise of his Mac suddenly seems to become irritating, because he smacks it shut and puts it away. Marco Iacoboni summarizes his approach.

"You can look at it this way. Neuromarketing is not so much a field of research as a playground. A playground where you can do things in a very unrestricted way and maybe encounter something that brings important new understanding. So you can then go out and test this understanding in classical academic experiments."

Hunting for *cool*

The word playground strikes me again when I'm at Caltech later – this time to meet one of Colin Camerer's professional colleagues, Steven Quartz, who is also flirting with neuromarketing. He's asked me to meet him at an outdoor cafe on campus, and it feels a bit like a play period. Small groups of students are swilling down lattes, as they swarm with the usual noise and tumult under the large canvas parasols.

There is also a playful quality about Quartz himself. Or, rather, *cool* is probably the best term. In my bag, I've got an old copy of the *Los Angeles Times*, with his portrait on the front page, looking just like Sting in his younger days. The forty-five-year-old professor has bleached hair, whose slightly darker roots are precisely dark enough to be fashionable and not just sloppy. His clothes are appropriately relaxed and smart enough to stand out from the crowd but in a cool way. Jeans and a tight flowery shirt that looks expensive. My understanding is that Quartz doesn't live in sleepy Pasadena but in trendy Malibu, where he is said to live next door to rock musicians and actors. And as a philosopher and neuroscientist, Quartz is the epitome of academic trendiness. For him, it's not boring credit cards or banal logos that grab the attention but film and designer products.

"My big interest is how the brain represents *value*," confides Quartz, emphasizing the word with a slightly clenched right hand. I notice the stunning watch on his wrist but can't identify the make at a distance.

"I would like to know how the brain enables us to navigate and figure out what is of value in our environment. And how it learns to make predictions about what yields a reward. I mean, one of the great watersheds of human development was the brain's ability not just to recognize value in the form of utility – rocks that could be axes, and plants that could be used as food – but also in the form of social value."

"You mean we use products in a social way?"

"Clearly. Funnily enough, this element has been lacking in modern economics, but even back in the 1700s, Adam Smith wrote about how economics is a social activity. People don't just participate in economic transactions in order to fulfill their basic needs but use the market socially. As a way of identifying yourself but also as a way of showing that you belong to a particular group."

The cluster of students behind us gets up all at once and wander out in formation. I use the occasion to follow up on Quartz's point, saying that people are known to rank themselves in terms of economic prowess. Studies in behavioral economics and happiness research show that we are more concerned with *relative* income than absolute income. It's not important what we have ourselves but how much we have in relation to our neighbor.

"Exactly – we're interested in the social value of objects. But how do these objects come to represent social value? How do things come to define and identify us? We still don't know how the brain does this."

Steve Quartz stresses to me that neuromarketing is only a sideline in his research, but it has nevertheless given him quite a lot of buzz. In 2005, *Seed* – a glossy, trendy American magazine on culture and science – crowned him as one of the year's icons in science. The specific achievement was the same experiment that put him on the front page of the *L.A. Times*, namely, the so-called cool scan, about which Quartz is also publishing a book.

"Can we come back to that a little later? Let me just throw out a few general things. When you are thinking along the lines of how the brain determines value, marketing and advertising are the areas in which everything plays out. Today, the array of goods is not a limiting factor. As Daniel Bell said way back in 1970 in his essay "The Cultural Contradiction of Capitalism," we are living for the first time in an economy that is not created to feed the belly but our lifestyle. Functionally, modern products are uniform. They do the same thing. So, what makes them different is how we use them socially. Take a phenomenon like a brand – what is a brand? It is a social distinction that we are creating, since there is no difference in the product."

"The essence of Montague's cola experiment, in other words?"

"Yes. Cola is brown sugar-water about which the brain discerns no particular difference until the brand information comes in. Then, the brain suddenly perceives an enormous difference."

As mentioned earlier, Steve Quartz has developed – and successfully marketed – a tool for studying the effect of trailers on an audience with MRI scanning, but he wrinkles his brow when it is suggested that neuromarketing might be something of a suspect area, in which trendy researchers try to make a quick buck. No, it's an area in which neuroscience naturally belongs, because it can answer some interesting questions.

"For me, it's more about the fundamental research questions than true commercial application. We're talking about relationships that define contemporary economic life. The market is the place where we exchange all these social signals with each other, and we really understand very little about what is going on at the neural level. My experiments ultimately have to do with studying how people assess objects socially."

"Like, for example, the cool-experiments," I say, so we don't forget them. "Yes, yes, that's clear," says Steve Quartz, nodding. "They try to pin down that indefinable quality of something being cool."

The experiment, which Quartz ran with Swedish designer Anette Asp, is pretty simple in and of itself. A group of people – women and men, young and old – were asked to take a position on a series of designer products and also on a series of famous people from the cultural arena. The question was whether there was some sort of common factor that applies to our judgment of things and people.

Practically speaking, it went like this: a group of trend-sensitive design students selected just under 150 well-known objects that they judged to have different degrees of cool. Their choices included a Louis Vuitton bag, diverse Prada artifacts, an Aeron office chair, an iPod, an ordinary Ford Escort and an array of other recognizable products. In the same way, they created a list of famous people with different profiles. Everyone from Uma Thurman and Al Pacino to gay icon Barbra Streisand and crooners like Michael Bolton and Barry Manilow.

In a fourteen-page questionnaire, the research subjects were asked to rate the objects and people on a cool scale from one to five and, when they were done, they took a trip through the scanner to look at some pictures. All the objects and people were thrown up on the goggle screens, one by one, while the scientists measured the brain activity they produced. Now you might think that it would come in the form of simple feelings of disgust or appreciation and prove to be in classic emotional areas, such as the ventral striatum with its dopamine and the good, old amygdala. But no. Things were happening in more sophisticated circuits and regions that are associated with complex phenomena such as self-evaluation, self-representation and identity – parts of the medial prefrontal cortex, which was also heard from in the cola experiment, and especially Brodmann's area 10, as it is so unromantically named.

"This fits well with the idea that the individual product has to be incorporated in some way into your social self," says Quartz. "So when you are making assessments, you're thinking of yourself in social situations with the product and of how it influences your status and other people's view of you."

Apparently, celebrities have a sort of product status in our brains. At any rate, the common denominator was activity in the Brodmann area. The cooler the item or individual was considered to be, the more intense the activity.

"In some people," says Quartz. "But after about twenty people had been through, we saw an interesting pattern in what provoked the activity. The surprising thing was that there were two big groups. One that identified strongly with the high ranking objects and people and one that, by contrast, reacted much more strongly to the objects and people they assessed negatively. Their Brodmann area really fired up when they saw a pair of hopeless shoes or a retouched photograph of Barry Manilow."

Quartz and Asp gave the latter group the designation Uncool, while the contrasting group was dubbed Cool Fools. The researchers have a theory that the two groups are driven by something very different in their relationship to products. Members of Uncool seem to be driven by a fear of seeming awkward and uncool, whereas the Cool Fools are driven by a desire to be cool.

"Of course, there are those who fall in between and don't react so strongly. Presumably people who don't take things so seriously," says Quartz.

"But what about the mirror neurons in all these people. Was there any activity there?"

"No, nothing really significant."

"I've just spoken with Marco Iacoboni, and he sees the mirror neurons as central to identification. Don't your different results say something about how difficult this is to interpret?"

His blue eyes wander for a moment, and then Steve Quartz replies, "You have to concede that. In the few published articles, you also see that it's a long way from the results to the discussion and the conclusions. There can typically be many different interpretations of the results. One contributing factor is that a lot of brain structures are participating in several functions and that they can sometimes work together on abstract phenomena, which can be difficult to get a handle on. Take the amygdala as an example. For a long time, it was associated solely with fear, because it always lit up in fear experiments. But later they found out that it can also be involved in pleasure and, today, they think that the amygdala probably deals with mood oscillations in general. It becomes active when stimuli are either positive or negative and, thus, *not* neutral."

Steve Quartz leans back in his patio chair. "But these are all just first generation experiments ..."

Another element in the coolness experiments involves looking at how the very different reaction patterns in the brain reflect differences in personality. There are studies that have shown that some people's brains are more receptive to positive information, while others are triggered more easily by something negative. People talk about positive and negative priming. Quartz and his people have put their research subjects through thorough personality tests and are now analyzing whether there are recurring patterns and connections.

It might be, for example, that introverted neurotics react with fear when they are exposed to Barry Manilow or uncool consumer goods. Likewise, it is possible that fashionable cool fools are to be found among the world's extroverted, careless temperaments. You could even begin to study whether the individual's attitude can be changed and, in the given case, how. The answers will go into a long-planned book for the general market, but hopefully some scientific articles will come out of it, says Quartz, even though he admits that there are some challenges to the latter.

"There is a culture that dictates that, if research touches something that has to do with everyday experience, it makes it almost unclean for science. But I seriously believe that this type of research is important for our economic self-understanding."

"And it doesn't worry you that you, as a scientist, are reaching out to people who just want to fob some goods off on us?"

"I couldn't care less about the sale of products. My primary motivation is to understand the *process*. But you know what? If these studies ultimately wind up being commercially interesting, I don't think it will be in the ordinary marketing of soap powder or dog collars. It will probably be in design. Think about it. How do you make products that are engaging to people? What is good design? It's all about perception and about experience with interacting with objects, and that is something you can delve into with neuroscience."

"But is this something that can grow out of your studies?"

"Maybe. At any rate, we've had cooperation with people at one of the big US design schools and we have discussed how neuroscience could be incorporated into design. The designers were very open."

Steve Quartz runs a hand slowly through his hair and squints his eyes against the sun, which is creeping in under our parasol.

"I predict the next big breakthrough will come when the field has become more integrated with social neuroscience," he proclaims. And it strikes me that this idea is yet another sign that the individualistic wave has just about peaked. In Europe, ad people and trend analysts are talking about signs of a turn in the conspicuous consumption society. We realize that unvarnished materialism and the intensive cultivation of self don't bring happiness and we are on our way back to something more community-oriented, a feeling that "we're all in this together."

"I mean, we had the 1960s and 1970s, when the counterculture was strong and when the collective meant something and people believed in it. Then came the 1980s and 1990s, when people worshiped the individual and were ironic. Now people are slowly progressing to a point where it's okay to be more serous and socially responsible, and I think this comes to some degree from the fact that we can see an ecological crisis around the corner. But it's also in our nature. The brain has some deep mechanisms for seeking affiliations and communities."

"Can doing neuroscience on the social aspects of our habitus play a role in getting us to change?"

"I think it's possible, if that kind of research can help articulate our social needs and get people to understand them better. Specifically, for example, the next step is that the research I'm talking about can help businesses make products that don't just satisfy an immediate need but also play a role in binding people together."

"How?"

"One example I see all the time is the Prius car, and also those yellow Lance Armstrong wristbands. Of course we know artifacts all the way back from prehistoric times, when there were emblems that held people together in communities and had a social meaning. Now, it's about to come back in the form of products, and it will be much stronger."

"So the Prius with its environmentally-friendly profile creates a sort of community? It's a tribal car?"

"Yes, but that's the way it is. A Prius owner always tries to tell you about his car, and you sense that he wants to win you over to the greater 'cause'."

"The gospel of Prius. Isn't there a parallel to the religious here?"

"There is. In the secular world, products can draw on exactly the same feelings and the same mechanisms."

"Your identity and your feelings take up residence in your things?"

"In some sense, we are what we buy, right? And if this identity can be about some deeper morality, some fundamental values, it will be even more powerful."

As I slurp down the last of my iced cappuccino beneath the umbrella, Steve Quartz saunters away down the winding paths of the campus to his office. I think how appropriate it is that he was named a science icon. The bleached philosopher *is* a sort of messenger of new times. Both Quartz and the smiling Iacoboni in Los Angeles are living illustrations of how natural science, which once stayed in its university confines, is now expanding its territory and brazenly moving into society's other sectors.

The ivory tower is abandoned and empty, as the researchers ally themselves with business and industry. Of course, they've done this for a long time, but you can see the outline of a new model for collaboration. It's no longer about the more traditional production line, where researchers come up with new knowledge and then sell it to the highest bidder. What is happening now is that the production of knowledge is becoming fully integrated into the reality of the end user right from the beginning. Brain research is a perfect example. Here, all kinds of interested parties are ready to propose experiments and even define new areas of research.

Marketing and branding are out there on the cutting edge, but the world of management is not far behind. Just a hundred meters from the building where Marco Iacoboni is "playing" with neuromarketing at UCLA, psychiatrist Jeffrey Schwartz is working with what he calls neuromanagement. Schwartz first became known for developing cognitive treatment methods for people plagued by the compulsive thoughts and actions of OCD, and his work on trying to change the thought patterns of these people sent him into the arms of the consulting business. The psychiatrist realized that management is also about influencing thought patterns, and that it could benefit from some cognitive input.

The philosophy behind neuromanagement is that businesses and organizations must address the challenges of the future with the methods of neuroscience. It is no longer business school philosophy, but knowledge about brain functions that will answer questions about how positive changes are to be encouraged. Insight into the brain's processes will show how employee education can be accelerated, how you can lead in uncertain times, and how you manage a flood of information to make better decisions.

Slowly, things are getting going. In May 2007, the first global summit for *NeuroLeadership* took place in Asolo, Italy. Here, while enjoying good food and the spectacular northern Italian landscape, you could also bask in the company of "an exclusive group of eminent thinkers, researchers and business people," and with them witness "the birth of a new discipline and the creation of new connections that will define new research, new tools and new ways of thinking about management."

New, new, new and everything centered around the brain. If there is anything all these germinating fields with the prefix "neuro" show us, it is the way to the neurosociety, a realm where human beings across all manner of fields and contexts are quite simply represented by their brain. It's the coming of the brain as the great common denominator in our thinking.

8

LIES, DAMN LIES — THE PRINTS ARE ALL OVER YOUR CORTEX

Where there are people, there are lies. As I mentioned earlier, the theory of Machiavellian intelligence claims that our capacity to deceive was developed by virtue of our distant ancestors' way of life and refined as their primate brains grew and developed more complex structures. Our closest relatives indicate that, from an evolutionary point of view, it has to do with the youngest part of the brain; that outer layer of coiling tissue called the neocortex, which takes up nearly eighty percent of human brain volume. The Scottish primatologist Richard Byrne and his partner Nadia Corp of the University of St. Andrews have explored the brains and behavior of eighteen species of primates, and they found a striking connection. The larger the animal's neocortex, the better they were at deceiving their fellow primates in everyday situations.

Homo sapiens lies all the time. As individuals, we discover the nature of the lie at around the age of three or four and, from then on, it is a natural companion without which only very few can imagine living. You can't really conceive of a modern, well-functioning society without the lie. Deception, of course, has a bad reputation, but in practice it functions as a sort of social lubricant that keeps the huge social machinery with its many cogs and their mutual relationships running. If we only told the truth, the whole truth and nothing but the truth at work or at home, we would very quickly have no job and no family. If politicians actually said what they thought and honestly laid out their ambitions and plans for power, they would never get it.

Language itself reflects the decisive role of the lie. Whereas there is practically only one term for truth, the lie has many names – some 112 synonyms in English. This is due, of course, to the fact that the lie has an abundance of nuances and facets, from awful, black lies to little white ones. And in the tension between the two poles, there are countless shades of grey.

There is hardly any religion that sanctions lying but if you read the Bible, for example, you can see an understanding of how fundamental and deeply human it is. It makes its entrance in the story of Creation, when Adam and Eve, as the story goes, lie to their Creator as soon as they have tasted from the tree of knowledge. Nor was their son Cain reticent when he killed his much-beloved little brother Abel. Yahweh came by and wanted to know where the boy was. "I don't know," replied Cain, "am I my brother's keeper?" In both instances, the lie was punished. Adam and Eve were banished from Paradise to a miserable earthly existence, and Cain was sentenced to a life as an outcast among strangers. God had seen all.

It is this trick some now hope to carry out by means of brain scanning. On the premise that "you may lie, but your brain doesn't," a select group of researchers are working on making MRI and EEG technology into a reliable lie detector. In the US, two commercial enterprises are already vying for the market. There is interest for their product abroad, and the debate is raging among lawyers and judges about the extent to which it is good or bad to look directly into the workshop where the lie is forged.

Finding a sure-fire method for revealing lies and deceit is, of course, an age-old dream. And all cultures have had their own traditions and folklore on how to identify the perpetrators. In medieval Europe, for example, people accused of witchcraft were tied into a bag and thrown into deep water – if they sank and drowned, they were innocent; if they floated, they were guilty and had to be burned immediately. In ancient China, people used to fill a suspect's mouth with grains of rice. They had to keep it in their mouth while the prosecutor read the charges out loud. If the rice was still dry after the accusation was read, the person in question was innocent, since the production of saliva was considered an expression of anxiety. People also placed great weight on nervousness in West Africa, where suspects had to throw an egg between one another. The person who dropped the egg did so because of being revealed and was, thus, considered guilty.

Pretty much the same principle is behind the apparatus we know as the lie detector, the so-called polygraph. The polygraph registers a series of physiological parameters that change involuntarily when we are nervous or stressed – blood pressure, respiration, heart beat and skin conductivity, which reflects sweat production, all of which are controlled by the autonomous nervous system. In a polygraph test, the subject is asked a broad range of questions of which some are neutral while others refer to the topic about which one wants to test knowledge. Thereafter, the graphs of the physiological parameters are analyzed and interpreted by experienced analysts.

The polygraph, which is an American invention dating back to 1913, is not used much in Europe. In many countries, it is not deemed to provide reliable evidence. In the US, on the other hand, it is used frequently by defense attorneys, prosecutors and police and, according to a Supreme Court decision, it is up to the individual judge whether polygraph data may be used as evidence in a case. The machine is often used in child custody cases, and it is almost always used when criminals convicted on sex-related charges are due to be released or paroled. This has apparently inspired the British. At any rate, in 2005, the British government introduced a law making it obligatory for someone convicted as a pedophile to take a lie detector test before they can be released on parole.

The lie detector, however, is under intense attack as ineffectual. There are even organizations – for example Anti-Polygraph.org – that are campaigning to scrap the apparatus altogether because of its lack of scientific grounding. The opponents claim that, if it registers signs of nervousness, it might just as easily be nervousness about the test itself as guilty knowledge. Moreover, people can learn to conceal these signs, even if guilty. There are a number of well-known cases in which high-profile criminals – from spies to serial killers – have passed lie detector tests with flying colors and continued their careers a bit longer.

The polygraph measures the body's inner environment, but there are also persistent attempts to read deceit on the surface. Psychologist Paul Ekman, who is now professor emeritus at the University of California in San Francisco, spent most of his career gaining expertise on how lying is reflected in facial expressions. He developed a Facial Action Coding System, which categorizes thousands of nuanced facial expressions that can be created by combinations of forty-three independent facial muscles. According to Ekman, we are not able to control what he called microexpressions, facial expressions that appear for less than half a second and which reveal a contradiction between what we say and what we feel. If you read these expressions according to Ekman's code system and compare them with body language in general, you should be able to discover deceit even at a distance.

Ekman himself claims that his system can be learned by anyone – the professor holds $35,000 five-day workshops – and that, after thorough training, you can spot other people's lies with ninety-five percent certainty. The idea and the course have been bought over the years by American ambassadors, intelligence services and police stations, which train their personnel to identify and be on guard against potential liars.

But both the polygraph and reading faces are external solutions. With the new scanning methods, there has come a hope of getting beyond these indirect measures to the source of the lie itself, namely, the brain. And it must be said that the movement has been surprisingly vigorous. The first feeble experiments were done around the turn of the millennium by psychiatrist Daniel Langleben, who was then affiliated with Stanford University, in the process of studying how certain types of drugs affected the brains of hyperactive children. By chance, during his work, he ran into the theory that one side effect of their disorder was that these children had difficulty lying. He was sure that the specific kids he knew could lie perfectly well. However, they might sometimes have trouble keeping the truth under wraps, creating problems for themselves in social contexts. Langleben got the idea of looking more closely at the lie's various stages of development.

His theory was that, in order to tell a lie, we have to undertake several independent mental operations. On one hand, the brain has to prevent the truth from slipping out and, on the other, it has to construct the lie itself and serve it to the world in place of the truth. Langleben believed that you had to be able to observe this dual book-keeping in a brain scanning as activity in various circuits. The lie, in other words, had to leave a physiological trace behind.

To test the idea, he didn't call in hyperactive children but ordinary university students, whom he instructed to lie about a particular playing card. They were given the five of clubs in an envelope and then went into an MRI scanner, where they were supposed to push a button "yes" or "no" to indicate a match, in response to a series of playing cards displayed, one by one, on a screen. The inducement was that they would win twenty dollars, if they lied so convincingly that the machine couldn't catch them. But as demonstrated in the article that was later published,[42] the students weren't very good at it. Even their innocent lie about a playing card left a clear imprint.

What immediately stood out for the researchers was that the lie showed increased activity in the whole prefrontal cortex; an indication that there was more thinking activity and cognitive work to lying than telling the truth. There were also special regions that stood out. The researchers put particular emphasis on the anterior cingulate cortex, whose function is still being debated but presumably plays a role when we deal with conflicting information. At any rate, it becomes highly active in the classic Stroop test, where people are presented with a series of words that describe one colour but are printed in another. When the research subject is asked to say what colour the ink is, they often choke and say instead the word that's written and, while they do it, their anterior cingulate cortex rings the alarm bell.

Daniel Langleben was hooked, and when he moved to the University of Pennsylvania, he continued his work with modified lie scenarios. In his second experiment, the participants could actively choose whether they wanted to lie to the nice researchers.

At a general level, the results were comparable – parts of the brain revealed that lying took place – but it was less clear whether you could talk about a definitive mapping of the lie's anatomy.

In the meantime researchers in Great Britain and Hong Kong produced their own independent studies of the phenomenon, and they all showed a clear difference between a lie and the truth in the individual research subject. But the experiments themselves were slightly different in their conceptualization, and there were likewise differences in exactly which areas of the brain were activated during the exercises.

At Harvard, psychologist Stephen Kosslyn began thinking about the matter, and he concluded that the provisional hypothesis was far too simple to describe reality. Lies and deceit can take infinitely many forms, and it does not necessarily seem obvious that the brain should treat them all in the same way. It feels different, for example, to come up with something on the spot as opposed to delivering a carefully considered falsehood.

And this is where Kosslyn struck. He wanted to directly compare the results from experiments with spontaneous lies *à la* Langleben and rehearsed stories that his research subjects had plenty of time to learn. They came to the experiment with their own account of a vacation about which Kosslyn asked them to change certain points. He suggested, perhaps, having the vacation take place somewhere else or coming up with a fictive companion. After a few hours of repetition, all twenty voluntary liars were put into the scanner, where they answered questions about their vacation experiences.

Totally in keeping with Kosslyn's suspicions, there were differences between the two types of lie.[43] For one thing, it was clear that the rehearsed lie did not involve the anterior cingulate cortex as much as a spontaneous deception. At the same time, there were striking differences in how memory resources were involved. Different parts of the frontal cortex came into play in the two scenarios, and the patterns fit very well with that of a person drawing on different types of memory. When a person lied spontaneously, there was activity in areas connected with what is called working memory, once labeled short-term memory; that is, structures and processes that temporarily allow us to fix and manipulate information. A sort of mental RAM. In the rehearsed lie, on the other hand, people drew on episodic memory, which corresponded to getting real recollections about situations from an internal warehouse.

Seen through Stephen Kosslyn's prism, researchers have only scratched the outermost frosting on the cake of deceit, and he has argued that an understanding of the phenomenon requires far more intensive research, attacking the problem from many angles. He says that to gain genuine insight into the mechanics of the lie will require us to delve deeper into fundamental phenomena such as memory and sensation.

However, while Kosslyn was thinking in complex terms, the growing interest in lies was no longer merely academic. At the American Defense Department, the field had been acutely deprioritized, almost hidden away, for years, but then the agency had its eyes opened in the wake of the catastrophe of 9/11. Now it was suddenly of vital importance to be able to evaluate the credibility of sources and statements, but unfortunately they lacked the equipment for it. This point was driven home in 2003, when the National Research Council issued its report, *The Polygraph and Lie Detection.* In this report, a series of independent researchers went through decades of lie detector data from, for example, the FBI, and the experts concluded that the polygraph was poorly suited to its job. The researchers likewise called attention to the fact the intelligence services, which were so busy using the polygraph, hadn't been able to develop a scientific basis for any sort of physiological lie detection method. And as the head of the investigation, Stephen Fienberg, said of the polygraph: "National security is too important to be left to such a blunt instrument."

But good advice is expensive. And, among others, the Department of Defense Polygraph Institute (DoDPI) had to pay up. After this depressing report, the nation's researchers were asked to come up with suggestions for projects having to do with lie detection, and it was the DoDPI's assumption that the brain was the way forward. The Department of Homeland Security and the Defense Advanced Research Projects Agency (Darpa) have followed suit and loosened their purse strings for research on lying. Darpa, for example, has given support to Daniel Langleben, who, after his first studies, quickly decided to get rid of the academic hair-splitting and get down to developing a functional lie detector.

With the experiences of the polygraph in mind, one of his goals was to create a format that was free of subjective interpretation. A data processor that is, so to speak, untouched by human hands. He realized this by developing a set of algorithms that could determine when the person in the scanner was lying or telling the truth. Andrew Kozel of the Medical University of South Carolina came up with the same idea and in 2003 published his own algorithms that could distinguish lies from truth in controlled experiments. Both instances involved computer programs that could analyze the data on brain activity from MRI scanners without human mediation and point out when the activity indicated a lie.

Along the way, the partial results hit the media as small stories, but 2006 became the year in which the topic became a serious subject of features and reports in the media. The big American newspapers and magazines competed to report from the front lines and run their journalists through the scanners of the research groups involved. What especially aroused curiosity was the news that there were two private companies that were tripping over each other to get a product onto the market first. Patent rights and technology from Andrew Kozel ended up with the Cephos company, while Langleben's know-how and algorithms were purchased by an outfit with the attention-grabbing name No Lie MRI.

According to the DoDPI, there are currently about fifty laboratories in the US alone working in one way or another on understanding and detecting lies – using not only fMRI technology but different forms of EEG, for example. Recently, psychologist Jennifer Vendemia at the University of South Carolina tested – with five million dollars from the DoDPI – almost seven hundred students with EEG measurements. She put them in a hood containing 128 electrodes placed to cover their scalp, and registered the electrical charges that came in the form of diverse brain waves. What interests Vendemia about the lie are so-called event-related potentials, ERPs, which are brain waves unleashed by particular stimuli.

If, for example, you show a person something visual, that person will, after 300–400 milliseconds, register a nice ERP, as an expression that "something" is happening in the brain such as a thought or increased attention. The EEG, unlike scanning, does not provide good spatial information, but it is much better at providing information about time. A functional MRI scanning only provides a picture every other second, while the electrodes on the scalp can register changes down to a thousandth of a second.

And it is in these time differences that Jennifer Vendemia's results may be found. In her experiments, she typically presents the research subject with some short statements that are either true or false and coded in two colors. Every time the statement is red, the person is to answer "true," while the required answer is "false," when the little sentence is printed in blue. However, the cards are set up in a way so that both red and blue statements can either be true or false. The research subjects answer as they should, but their ERP pattern reveals that they are stating a lie. If you get a red statement "A snake has thirteen legs" and answer "true," it takes 200 milliseconds longer to answer, and the ERP signal is stronger in regions in the middle and top part of the head. Roughly speaking, the MRI experiments also point toward some of these areas.

Jennifer Vendemia doesn't think her method can be hoodwinked by good liars. At any rate, she has looked at the extent to which practice can change the extra reaction time and found that even trained liars have exactly the same ERP pattern as pure novices.[44] Her most interesting claim, though, is that her measurements can predict a lie *before* the liar has decided on it. Thus, she sees the first changes in the person's EEG about 250 milliseconds after the statement appears on the computer screen, while it takes between 400 and 600 milliseconds before the pattern showing a decision appears.

There is a third player up in Seattle, Lawrence Farwell, whose Brain Fingerprinting Laboratories is marketing his own Brain

Fingerprinting method. Like Vendemia, it is an EEG technology, but Farwell uses an electrode-studded headband instead of a hood, and he concentrates on the so-called p300 brain wave, which is a part of the overall ERP pattern. What he is testing is the extent to which a person recognizes a given stimulus. It can be anything from a telephone number to a picture of a decrepit summer house in the country. And the principle is that something you have seen before unleashes a characteristic electrical response between 300 and 800 milliseconds after it is presented. In EEG readings, you can see the p300 wave as a clear peak on a curve with smaller waves and, with his own patented algorithm, Farwell believes he can detect a lie with nothing less than one hundred percent certainty. In his sales materials, he states that he has tested the technique on two hundred research subjects in projects financed by the CIA and the FBI; but, in published articles, it is down to six research subjects.[45]

Farwell and his headband have appeared on all the big TV stations in the US, but Brain Fingerprinting has also had its debut in the courtroom. In 2000, a District Court in Iowa conducted a hearing to determine whether Terry Harrington, who had been convicted of murder in 1978, could have his case reopened. Hired by the defense attorney, Farwell tested Harrington by showing him pictures of the murder site, and he testified that the convicted man had never seen it before, according to his p300 results. Then the only witness in the case admitted to having lied about seeing Harrington at the murder site, and the judgment was ultimately reversed. In connection with the hearing, there was an eight-hour long discussion about the extent to which Brain Fingerprinting could be admitted in court and, in 2001, the judge determined that the test lived up to the legal standards for scientific evidence.

Some also believe there are signs that MRI technology is not far from being admissible. At any rate, in 2005 the US Supreme Court determined against the execution of minors, a decision partially grounded on MRI studies that show that, in many regions, the brains of young people do not function like hardened adult brains.

Yet, even as MRI-driven lying research has been transformed into product development and marketing in record time, there are a few worriers on the sidelines shouting concerns. Sociologist Paul Root Wolpe, a professor of bioethics at Emory University, imagines a violent counter-reaction from the general public. Wolpe believes that many will quite simply see the technology as a piece of creepy science fiction reminiscent of mind-reading and surveillance *à la* Orwell's *1984*. He speaks indefatigably for an open, popular debate on the subject.

The American Civil Liberties Union, the ACLU, agrees. They are particularly nervous that a new, sexy technology to detect lies will be misused under the aegis of the war on terror. In the spring of 2006, the ACLU held a symposium at Stanford University, at which selected researchers, philosophers and other observers gave their interpretation of the matter, and subsequently they asked for access to records from the American government. They wanted to give the public an insight into how huge the funding is for research into MRI and other lie detector technologies.

Even in Europe, we've heard an echo of this discussion. In 2005, a few hundred invited citizens and neuroscientists met at the Meeting of Minds conference in Brussels, where they discussed the future significance of brain research and knowledge about the brain. One of the problems discussed was the possibility of "mind-reading." Specifically, Brain Fingerprinting was discussed, and several prominent scientists expressed fundamental ethical concerns about the potential for this type of technology to invade our inner space. When you open up the brain's processes in this way, you violate, in a heavy-handed way, the individual's right to keep his or her thoughts and feelings private.

Guantánamo prisoners and CIA agents

At Cephos, people aren't nervous but rather expectant. "Cephos is founded on the simple premise that truth is a valuable commodity," it says on their homepage. The man behind the idea – today both CEO and majority shareholder – is thirty-five-year-old Steven Laken. A man who is known in scientific circles as something of a wonder boy. As a twenty-six-year-old Ph.D. student, he was a rising star at Johns Hopkins University, where he developed the first blood test for hereditary colorectal cancer – an achievement that resulted in profiles and reports in everything from the *New York Times* and the *Wall Street Journal* to the major American talk shows.

Today, Laken runs Cephos from a couple of anonymous offices in Peperell, Massachusetts, but he would rather meet at the garden of the Boston Public Library. "It's just so wonderful there," he states in his e-mail. And now I'm here, on the lookout between patriotic friezes and statues of lions standing as memorial to the Civil War fallen. I wonder how I'm going to pick out Laken. A stream of people passes through the large, open gates of the main entrance, and I have no idea how to spot a man who makes his living sniffing out lies.

He doesn't look like a scientist, it turns out. Steven Laken – it's finally he who finds me – looks like a young businessman. Nice blue suit, black briefcase and short blond hair, carefully groomed. Laken tells me that he's from the Midwest and has Norwegian and Swedish roots. He also has kind, light-blue eyes, and he insists on buying me a cup of coffee. We sit down in the library's central garden, recently renovated but only sparsely populated and pleasantly screened from the hubbub of downtown Boston. Unfortunately, there is also a fountain that makes an incredible racket. It competes so much with Laken's somewhat muffled, reticent voice that I practically have to crawl across the little café table between us to hear him.

Perched like that, I confess that I've wondered a lot about how a person can leave a position as a hot-shot researcher specializing in the molecular biology of cancer to peddle high-technology lie detectors.

"Excuse me, but it just seems strange," I say. Laken smiles warmly. Almost indulgently. He had always – even when he was a student – wanted to start a business, he says. And when he developed his sensational cancer test, he got involved with a small enterprise called *Exact Sciences*, which introduced the test to the market and then went public, making him a huge pile of money. It was sitting in the bank, just waiting for some action.

"Then came September 11, 2001. Afterward I thought a lot about the people we have locked up in various places in the world – Guantánamo, Afghanistan – and whether there were ways of finding out whether they had information or not."

The young researcher-businessman had also followed what was going on in neuroscience. "Those beautiful colored pictures of the brain lured me in," he says. As an indoctrinated molecular biologist, he thought at the beginning that it must be possible to identify some proteins or stress hormones that could be tested to reveal lies.

"But where do these hormones come from? The brain. So I reached the conclusion that we had to look directly at the brain."

He got hold of, among others, Langleben's and Kosslyn's first published studies and, in 2003, he sat for three months in Harvard library, looking for relevant research results and patents in the area. By "a stroke of pure luck," he also participated in a conference in New York, where a small group of researchers spent a whole week discussing scanning techniques and the mapping of the brain. There, he ran into Andrew Kozel. He had not yet published anything about lies, but he showed him his introductory work, and Laken could see that the researcher from South Carolina was on to something. He measured different people and different sorts of lies with different scanners and, across the gamut, he got some reproducible differences in the activity patterns. Laken could smell the commercial possibilities. He looked for some investors, convinced a couple of rich business angels to contribute some cash, his wife committed some money, and he got going.

"At the meeting in New York, I felt very intensely that neuroscience was the new genetics. It's right where genetics was fifteen years ago. And the scientists are making some of the same mistakes. They're creating a lot of hype about apparent connections that we don't really understand yet. It's just like when they began finding genes that seemed to be linked to particular diseases, but they had no idea how they caused the symptoms."

"But didn't you think about whether it was a good idea to get into this new area? Whether it might not be problematic to look for lies?"

He pauses, makes a face, and looks out at the fountain.

"At the beginning of the Manhattan project, when the goal was to produce an atomic bomb, there was someone who said that science always pushes the boundaries of ethics, and it's true. Ethics is seldom out ahead of science."

I can only agree.

"We're living in an age that is very different from the time before 2001," he says, leaning forward over the table. "But if what we're suggesting can be done at all, then you have to have a cooperative person. You can't force anyone who refuses to cooperate and just stick his head into the scanner. And when it comes to detainees, for me, it's all about making sure we do whatever we can to release those who are being held for no reason. It's clear to everybody that there are situations in which people are taken from their homes and their country and registered and mistreated for no reason. I would like to prevent that."

"How does the market look to you?"

"The market is phenomenal," says Laken, plopping his hands down in front of him. A sudden and awkward movement for the otherwise very controlled man. "In the US, every year, there are – and this is amazing – thirty million civil and criminal cases. In perhaps ten percent of them, large sums of money may be involved, and the whole case turns on he said/she said. Nobody can really decide who is telling the truth. Here, fMRI can come in."

"If it's approved."

"Yeah, well, approval would be great but it's not crucial. Look here. It's already the case that the polygraph is used 250,000–500,000 times a year, even though it's not accepted as evidence in a court case. It's used, for example, by defense attorneys to plant a message in the public at large that their clients are innocent. Did you follow the case about the Duke lacrosse players accused of raping a stripper?"

"I've seen something about it."

"Okay, but they voluntarily took a polygraph test and passed with flying colors, and the message from the defense was: forget it, they're innocent. The 'court' of public opinion is very powerful, and as a defense attorney you try to influence the potential jury pool and public opinion at large. With MRI, you can say to the other side: look, we have an interesting new tool. Just think if we're allowed to use it. It's a negotiating tool."

"The public is also fascinated by those beautiful images of the brain and the idea that you can actually *see* our innermost thoughts."

"Clearly. But, of course, the court tries to distinguish between what looks like science and what *is* science. Neuroscientists at the moment agree that fMRI *can* do something with respect to lies, but they have to find out exactly what it is the data supports."

"And there is also a market beyond court cases?"

"Absolutely. The government uses the polygraph as a security check on people who are already employed as agents for the FBI and the CIA, or people who want to be employed there, and there are up towards half a million tests a year. Just background checks – you know, questions like: have you ever sold drugs? Have you ever used drugs?"

"Did you inhale?"

He laughs for the first time. And the smile disappears as quickly as it came.

"My calling here is that the US is at war, that there are still detainees, that *al-Qaeda* is still out there, and that the American and other governments are still fighting a war on terror. If I can help prevent another September 11 and make the world a safer place, I'll be happy. Even if I don't earn a dime."

He might very well though. Cephos is not in the market yet, but they are already talking with various agencies of the American government. The researchers at the University of South Carolina receive money from the Department of Defense, and the company collaborates with them on scientific studies and drawing up guidelines to be followed.

"They are incredibly interested. Today, they use polygraphs and interrogation, and even if they have now been allowed to do certain things ..."

"You're talking about enhanced interrogation procedures. What some people call torture?"

"Hmm, yeah. But the question is whether that's the right thing to do. If you have the choice between MRI scanning and some soldiers doing what they did at Abu Ghraib, I believe that MRI is the way forward."

This leads us to the technology itself. How much can it do, when it comes down to it? The researchers see different areas of the brain become active in their various experiments?

"Correct. It is very difficult to do a direct comparison from lab to lab. But there is a certain intersection. Most people see increased activity in *particular* areas – namely, the right medial frontal lobe, the right orbitofrontal lobe and the anterior cingulate cortex. Mostly on the right side. This is also the pattern we see.

But it depends on precisely what type of scenario you are using to test your lie. What we have tried to show is that, if you use selected areas as markers for lies and deceit, there will be greater activity in those areas nine out of ten times, when people consciously lie to a yes or no question."

Laken has brought his evidentiary material along, an article about his most comprehensive experiment to date.[46] Sixty-one research subjects were tasked with stealing a ring or a watch without the researchers knowing what they had swiped or whether they had taken anything at all. The researchers then set about determining whether the subjects were lying. I want to point out a passage in the discussion section: "This is the first study to use fMRI to detect deceit at the individual level. Additional work is required to clarify how well this technology will function in other situations and other populations."

Until now, all the lie detection has taken place as controlled experiments. Rings, watches and playing cards. But how will it work in the real world, where there can be all sorts of lies and you're only dealing with one person and not a group of volunteer students that you can analyze statistically?

"Nothing simpler," says Laken, shifting into his teaching mode. "The interested person contacts us and meets us with his lawyer. We sit down and look at the case and form a series of specific questions that have to do with the given crime. Some will be lies, some truth. It could be questions about whether they were at home, when the murder was committed, if that's what they claim in court. We put together about twenty yes / no questions that the person must understand and know about in advance. The scans themselves take place with Andrew Kozel in South Carolina. First, they practice and then are put into the scanner and get a randomly selected question every six seconds. They see them on a screen and answer them with a mouse. There are three sets of questions, where some are neutral – something like 'are you Danish?' or 'is it Tuesday today?' Every second and a half, the scanner takes a picture, and all the pictures taken over the space of a half hour are sent to us. We analyze them with a software program developed specially for this purpose in England."

All the pictures are translated into graphs and numbers – pure data – which takes about half an hour. Then the computer is asked what happens when the critical questions come up. Do the data look markedly different in relation to the other questions? It gets a chance to mull this over and comes up with an analysis – not a single living person involved – of what areas of the brain were active during these particular questions.

"Were you at home, weren't you at home, for example. The computer shows where there was the highest activity, and that's what we say is the lie. We look for what we call groups one, two, four in our article, that is, the anterior cingulate cortex, the right medial frontal cortex and the right orbitofrontal cortex. I just want to say that there is a lot of difference in the activity if you compare different people, so it's a matter of relative activity in the individual. I can't compare you and me, for example."

He can show me some pictures, though, and takes the computer he brought with him from his briefcase. The first thing he pulls up are some grainy, grey pictures of a brain.

"This is what comes out of the machine," explains Laken, punching some keys that make the pictures even more blurry. Almost dingy. But you can see the luminous spots of activity better against the background. He shows different research subjects when they answer whether they have taken a watch. There is activity in a lot of areas, but when the computer program cleans it up, the lie appears nicely in yellow and red. Small round splotches in the right places.

"Let's look at some more, just for fun," he says, and he entertains us with a whole suite. When I think I've seen enough, I ask whether they have tested some really good liars. The liars' elite. It's one thing to test those of us with average abilities for deceit, but you might presume that professional liars, who have trained and perfected their technique and style, can do something the rest of us can't. Just like with the polygraph.

"Of course, that's something we'll find out. Right now, in fact, we have a grant from the National Institutes of Mental Health to test children who are pathological liars. There's a group of really bad juvenile delinquents who have been diagnosed as being pathological liars and who simply lie all the time about everything."

"They can't help themselves?"

"No, apparently not. But what we're looking at is whether their brains react the same way as everyone else in our experiments. Okay, there are certain things that seem to be different – a region that has to do with impulse control lights up in these children and not in normal control subjects. This may have something to do with the fact that they have problems controlling their behavior. As yet, we've only studied three children, but the study will be expanded."

"But it looks like experienced liars are just like the rest of us in a scanner?"

"For now. We have also tested the concept on journalists – yeah, sorry – and other people who say they could fool the test, no problem, because they were such proficient liars. They couldn't. Their brains looked like everyone else's, even when they really try."

Laken and Kozel even tested a couple of subjects for whom English was a foreign language, and there was still no difference.

"Clearly, there is a difference between a rehearsed lie and a spontaneous lie. That's something Kosslyn over at Harvard has shown, but for our population, it's only rehearsed lies. You don't want to surprise anyone in a test like this, because you get all sorts of white noise that's hard to interpret."

Steven Laken sighs and a small wrinkle appears above his left eyebrow. At some point, he says, they need to start testing hard-core criminals to see how they do. He could imagine testing some inmates in cases where there is DNA evidence and you have eye-witness accounts supporting the verdict, so you know the right answer.

"But I want to emphasize that we are trying to use the technology in civil cases, where MRI data is just another piece of forensic evidence that can be put onto the scales. It takes more for criminal cases, where there can't be a shadow of a doubt. Our method is not like DNA, which provides pretty much 100 percent proof. I would never use an MRI to say 'here's evidence that someone is lying.' It's along the same lines as other forms of scientific evidence – handwriting analysis, fiber analysis, shoe prints, and that sort of thing."

Out in the American legal landscape, there are attorneys who are actively trying to get brain scanning accepted into the system; various cases in which the brain is in one way or another at the centre. Steven Laken himself mentions a man in a car accident that left him brain-damaged, so he couldn't continue in a high-paying job. With scanning evidence and affidavits from neurologists in hand, he was able to get compensation.

"Scientific evidence has to have a sound scientific basis and thus be reproducible and generally scientifically accepted," he says. And then: "We fall under those criteria."

"But you're still not offering your test yet?"

"No, because we want results that show more than ninety-five percent certainty for spotting a lie. We didn't get that in the study of sixty-one people, so we're doing another study. I want it to be *really* good before we go to court for the first time."

"But is there interest from anyone besides the media?"

"Lots. Just this morning, I got a call from someone accused of a crime he says he didn't commit. And we've had contacts from Australia, China, Mexico, Spain, Portugal and Germany. If you're accused of something you haven't done, you'll do anything – even fly to the US and subject yourself to an MRI scan."

And it's not even that expensive, Laken opines. The scanning itself only costs three hundred dollars, and with the development of a good set of questions and an evaluation of the case, the whole package comes to $5,000–$10,000.

"And that's not much more than a minor face-lift."

May it please the court

"Prosecutors, defense attorneys and accused would explode if they had lie detectors on in the court room," the Danish defense attorney Peter Hjørne has said, just a year before he was disbarred for leaking confidential information to the press. "There's lying going on all the time."

"I don't want to comment on a statement like that," says Robert Shapiro from his office in Los Angeles. He is one of the most notorious defense attorneys in the US, particularly after helping to acquit ex-football star O.J. Simpson of the murder of his ex-wife Nicole and her lover. Shapiro is also a board member for Cephos and a co-owner of the company. And he's a busy man, whose secretary has a hard time finding five minutes when he's in the office and can take a phone call. But she does, and Shapiro goes directly to the heart of the matter.

"I was contacted by Laken three years ago and could immediately see MRI as an extremely exciting and interesting tool. It's one of mankind's oldest dreams – to be able to uncover a lie. So, I've invested in Cephos, because I'm convinced there is a huge market – there has always been a market for testing whether people are lying, right?"

"Er, probably."

"But if you ask me, the definitive use will not be in the legal system but in the military, in intelligence services and the whole governmental apparatus. Security checks."

"What about attorneys like yourself?"

"For me, it's very relevant for internal use. I already use the polygraph in my practice."

"To avoid defending people who are guilty?"

"Because I want to know the truth for *myself* in my client's case, so I can help him or her to the greatest extent possible," says Shapiro in a tone indicating that I've said something completely idiotic. Then he picks up the thread again.

"But there are problems with the polygraph, and MRI scans are far more scientific. If the precision also gets better, there are immense possibilities for application."

"But will the method be recognized by courts?"

"I think it will be very difficult. Not because of the science but because the technology will take something decisive from the jury, the twelve people we have decided are to judge what is true and false. The *system* doesn't want it to be too machine-like, for justice to be completely taken out of human hands."

Shapiro isn't the first person to try to get the technology recognized and won't be the last, but he has no doubt that it "is so sexy that attorneys will throw themselves at it."

"Attempts will be made in one state after another but remember, it will only happen in cases suited for it. The case will have to be about something specific – did you steal *this* car or did you kill *this* person. Something black and white, in other words. When it comes to degrees of guilt, which it often is, you can't use it."

Who the pioneers will be Shapiro has not a moment's doubt. With a quick, almost monotone voice, he says, "Well-to-do individuals with suitable cases and aggressive lawyers."

"Yes, we have three categories of lawyer now," says David Faigman two minutes later on the telephone. "Those like Robert Shapiro, who are enthusiastic about MRI, the skeptics, whose skepticism derives from their experiences with the polygraph, and those who haven't discovered the technology yet."

Faigman himself is an academic lawyer. As a professor at the University of California, Hastings College of Law, he has researched and written about the use of scientific evidence in the American legal system. In contrast to his colleagues in private practice to the south in Los Angeles, he doesn't believe that there is anything in principle to prevent MRI from being used in court cases.

"The science behind it might be a little uncertain right now, and the technology will have to demonstrate that it functions reliably in particular uses. But if it looks promising and the research results start piling up, courts will allow MRI. First, for some things, then more later. For example, it might be hard not to allow it if an accused wants to use it, because for constitutional reasons people have to give the benefit of the doubt to the accused."

"Isn't it also hard, because it's a sexy science?"

"Sexiness won't affect the practice. I think. In part, courts are becoming ever more sophisticated in dealing with scientific evidence; in part, there is also a lot of anxiety around neuroscience."

Unless there is specific legislation on the new technology, it will be up to individual judges to weigh anxiety vs. interest and let it into their courtrooms. And it's still early yet, estimates Faigman.

"But if I had to guess what cases will be the first ones, I think it will be people already found guilty who are trying to re-open their cases. Death-row inmates who don't have DNA evidence but who claim that a lie detector will raise serious doubt about their guilt."

If the MRI lie detector proves its validity, Faigman thinks it will be a "fantastic tool." At least, when it comes to cases that can be boiled down to simple questions. But in contrast to Shapiro, he doesn't think it's suitable for testing hopeful agents for the CIA or FBI.

"Not at all. Here, you're looking for personality traits and you're trying to reveal personal tendencies that aren't desirable. And that sort of thing rarely comes out in questions you can answer yes or no to."

"But they're using the polygraph today?"

"That's correct, but I've just done a report that recommends they limit the use considerably."

"But don't the insecurity of the times and fear of terror mean that it will be tempting to use any technology that promises to uncover lies?"

"Well, maybe," says Faigman. And he immediately adds that there will undoubtedly be more pressure to use MRI technology in "every conceivable situation as it gradually improves."

"You can't see all that much in a scan today, but let's fast-forward to the year 2026 – by that time, you'll be able to interpret the colored spots far better and get far more information from them. That's what the whole of neuroscience research is all about – to increase our understanding of what our brain activity means. We can't get around the fact that it will raise questions about a brave, new world. *All* biological technologies, whether it's brain scanning, DNA or measurements of physiological data, are developing in a direction of revealing more and more about us. To be able to look directly into the brain and read things that for the whole of human history have been internal and absolutely inaccessible is very taboo-breaking."

"Is it about to redefine our idea of the private sphere? What we today consider our innermost, inviolable core will no longer necessarily be our private sphere."

"No, you're undoubtedly right. At least, not in every circumstance. And this makes for a huge challenge for lawmakers always to weigh when and where you can use, for example, MRI lie detection. We will also see some development there."

"Like what?"

"Today, MRI technology comes under polygraph legislation in this country, which means that it may not be used in connection with employment in the private sector or by insurance companies. But if you *can* actually reveal a lie or a swindle, it is possible that people at some point will ask why a business shouldn't have the right to use it."

Springtime for Stalin

At No Lie MRI, they are just about ready to offer the world protection against despicable liars. And as one gathers from the firm's homepage: "Liars benefit from their skills at the expense of other individuals or groups. It is estimated that at least one out of three conversations involve a lie, and that five percent of the population are what is referred to as 'seamless' liars or 'natural' liars."

Like Cephos, No Lie MRI is scattered in its geography. Founder and director Joel Huizenga has an office in San Diego, but he suggests we meet in Newport Beach, a small coastal town south of Los Angeles. He is contemplating leasing some MRI facilities here. The meeting place is a private clinic that specializes in imaging techniques for all sorts of medical purposes. It's an inviting building, painted a *café au lait* brown, whose parking lot is filled with quite a few equally inviting cars – a silver Audi TT, a few Porsches and a fire-engine red Mercedes sports car.

"The clinic is owned by two cardiologists," explains Huizenga, who arrives in a nondescript grey minivan. "Doctors are typically bad businessmen. There is scanning equipment in there worth ten million dollars that they don't have customers for!"

The large, bright lobby with soft cocoa-colored chairs is empty except for a team of smiling receptionists and an expectant couple with two small children.

"What did I tell you? No clients," gloats Huizenga, who indicates a shaded corner, where we can sink into two comfy armchairs. This tall, blond-haired man is in his mid-forties, and he explains proudly that his unusual family name is Frisian. I'm a little more interested in his professional background, and this turns out to consist of a bachelor's degree in molecular biology and a master's degree from business school. The business credentials shine through clearly but in a more direct way than in his East Coast competitor, Steven Laken. Still, just like Laken, Huizenga got into the lie-detecting industry by chance, when he heard about the new research and could immediately see the commercial possibilities.

"It was in 2001, and there was a little notice in the *New York Times* that Langleben had some preliminary results about detecting people's lies in the brain. This was extremely interesting, because I had already run a business using MRI to detect calcification in the cardiovascular system. With patents and completely automated software. So, I called Langleben and told him I could automate his lie detector. I talked with them and was given access to their patent and techniques."

"They were on the lookout for a commercial partner?"

"No, God, no! They had no idea that what they had could be used for anything. Even the university's office of technology transfer didn't believe it had any use whatsoever. Nor did they believe it until December 2005, when the article came out showing that it worked on individuals."

"It was around the exact time that the Cephos people came out with the same thing?"

"Yes, that's right. But their business foundation was just a single guy from South Carolina, and they have no automatization of the processes. They do it manually."

"Laken told me that the analysis takes place fully automatically with software developed in Great Britain."

Joel Huizenga looks mildly irritated but responds pleasantly.

"Well, Langleben's group from the University of Pennsylvania has published an article on their software; Cephos hasn't. And to be honest, I don't even think they can get their product to work. Everything they have done, they did later than us, including applying for a patent. So, I don't think they'll get the patent. And take note that they don't have any investors. It's just their own money. I'm certainly not nervous about the competition," says Huizenga, demonstrating his disregard with a condescending expression. Then he smiles slyly and says it might actually be good that Cephos exists.

"Journalists love to have two players in the market, so it doesn't look like a monopoly. The technology is scary enough as it is."

"Scary?"

"Yeah, the reporters all treat it like a horror story. It's those feelings that make the story so saleable."

"Couldn't it simply be that it's interesting?"

"Hah! *Interesting* is only for intellectuals. For the rest of the world, it's all about fear and greed."

He laughs a bit shrilly.

"People are curious, right, but also afraid of someone coming in and looking at what's inside your skull. The really scary thing about MRI is that it actually seems to work. As opposed to the polygraph."

He is quiet for a brief moment and then goes down a completely different path.

"You know, I hadn't thought much about lies before I got into all this, but now I see lies everywhere. You can't open up a newspaper without lies, deceit and non-disclosures flowing out. And in your everyday life – people lie to you all the time for all sorts of reasons."

"But MRI is hardly a tool for everyday, trivial lies?"

"Are you crazy?!"

Huizenga shakes his head and clucks his tongue, as he places his hand on my shoulder.

"What people are worried about in this world are sex, power and money in that order, and that's what they lie about. If you're talking about private life, sex, so far as I can see, is extremely important to people, and we know that every tenth child has a different father from the person he or she thinks. There is a lot of male anxiety and insecurity out there – is my wife faithful or does she step out on me? It could be a *huge* market for us."

Still with his hand on my shoulder, he says that No Lie MRI will actually go to the public with this sort of infidelity case. They will test and demonstrate their technology to an invited audience consisting of a camera crew from CBS and NBC. It will take place live and, hopefully, reach the entire nation.

"Our attorneys have told us that we have to select a case that's not criminal but will resonate in the media, and we've had a lot of women call us wanting to prove their fidelity to their husbands." He blinks at me. "Interestingly enough, there are no men who've wanted to do the same."

Then, he changes tone and explains that No Lie MRI is also working on more serious cases. Murder cases.

"We have a murder case on its way to court. But we keep that sort of thing behind closed doors. It's not the attorneys but the people themselves, for the most part, who contact us and we try to winnow out the cases that will present the technology in a good way for the legal system. We get the other clients to sign an affidavit that will make it difficult for them to use it in a court case, because we don't want to risk bad cases or incompetent lawyers who will smear the technology and affect our opportunities for using it scientifically or strategically. But to sum up, we're going after the personal market first, and then the legal, and after that, whatever comes up."

"What could that be?"

"Note that our whole society is based on confidence," says Huizenga, spreading his arms wide. "We have, for example, a huge market for Internet dating, have you tried that?"

I have, but don't get a chance to lie about it before he moves on to his point.

"People lie like crazy when they're trying to sell themselves. They present themselves as something completely different from what they are, and this means that the dating companies don't solve the problem for singles. You don't know who the other daters are and therefore you don't have the advantage that the companies claim to provide – namely, that you can seek out the person you want. Women send in pictures of their daughters, and stuff like that. Almost every third person on these big portals turns out to be married! It would be a valuable service for the daters if you as a provider could screen the users in some way and secure yourself against this sort of out-and-out lie."

I think of Steve Laken's comparison of the costs to the price of a face-lift and ask what No Lie MRI has thought it would charge for their services.

"I would think around thirty dollars a minute and then an initial fee that covers our work sitting with the customer and finding the right questions and putting them into the right computer format. If you take women who want to be cleared of suspicions of an affair, we're talking about nine hundred dollars. That's dirt cheap."

And Huizenga is not afraid that the media's horror scenario and people shouting about ethical concerns will stand in the way of his business. The ACLU's concerns and seminars don't impress him. At any rate, he doesn't reveal any trepidation.

"Actually, most of the people who have been talking about the drawbacks haven't even thought about the worst. What I see as the most terrifying perspectives are in the area of power politics. Imagine if the next Joe Stalin was in the offing somewhere and out to solidify his position by getting rid of his enemies. In his day, Stalin just killed everyone suspicious and, undoubtedly, got rid of a lot of people who *weren't* going to betray him. With a reliable lie detector, he would have been able to concentrate on those people who were actually disloyal and dangerous and have had a far more effective regime."

Joel Huizenga looks at me as if he's expecting a reaction, but his last sortie has completely thrown me. I just stare back.

"You think that's too far out? But are you sure you can't imagine Western governments who might want to test their officials before they take their posts? I know at any rate that the parties in the US are very clannish and want complete loyalty. And when I go on radio or television, this is the first topic that's discussed – always politicians and their lies. Politics and lies are inescapably linked."

I ask whether there is anything at all Joel Huizenga would not put his name or MRI scanner to. He purses his lips and seems to be mulling it over. Then he nimbly avoids an answer.

"I can imagine there will be a lot of things proposed that will prove to be unethical, but I can't guess in advance. So, we take it on a case by case basis. The way we're getting the business on its feet, it will be very anonymous and discreet for the client.

Contracts will be made with the various MRI centers, where the client comes in and is put in the machine by a technician who doesn't know anything about the person's case. The questions can be asked over the Internet, and the results will go directly from the scanner into our central computer and be analyzed there. These centers allow us to control what sort of business is done, and we won't test anybody who doesn't want to be tested. Also, clients can only be tested on the things to which they have consented in writing. If it's about whether they took money from the till, we won't ask them whether they've had an affair with their secretary. We don't give our software out, and we only work with centers that certify that they do things our way. We have enough business that we can do it in an ethically defensible way and still have plenty of clients."

To be honest, he's already had international opportunities, says Huizenga. At the moment, he has some hot contacts in Switzerland – there are just some legal matters to be taken care of, and then: "Who knows? The other day I was interviewed for an international journal for MRI specialists. Have you seen the article? I've got a great quote."

Huizenga looks out at the empty room and sketches a rectangle in the air with both hands. An imaginary neon sign, I suppose. Then he quotes himself solemnly:

"Coming soon to an MRI centre near you. Isn't that just great?"

Crime and Punishment

After being fed the grand visions of private enterprise, I want to check out what academia has to say and head to Stanford

University in Palo Alto to talk to law professor Hank Greely. He is at the forefront of a group of academics who are thinking about how modern biology affects the social, ethical and legal aspects of life.

"Come in, take a load off," says Greely, when I arrive – late and flustered. I've been desperately driving around the large campus and managed to park illegally twice, before I finally reached the right building. Greely, in a spacious office chair, is sitting like a calm, smiling Buddha. He looks like a big, grey-haired teddy bear, holding a can of diet cola in his paws.

"Have you met Huizenga," asks Greely with curiosity. "Is he a charlatan?"

"I think he's a guy who doesn't care how he makes his money, as long as he can see an interested market."

"Publicity hound?"

"Well, at any rate, he's called in two national networks to witness the debut of his product on the market."

Greely nods a few times and says that Steven Laken seems more credible to him. "I get the feeling he really *believes* in it. And you know what? If you *can* develop a reliable lie detector, I think there are a number of legitimate uses and things the technology can be used for."

His primary concern seems to be whether the technology actually *works*.

"When I first heard about this back in 2002, my reaction was: this is really interesting! I wonder if it works and, if it does, what do we do? That's still my reaction."

I'm a little surprised about how positive the neuroethicist Greely is about this newfangled lie detector. It is clear that, beneath the concerns about the extent to which it works, he actually sees MRI as a potentially beneficial contribution to the daily practice of law.

"Indeed! It could make a huge number of cases disappear, right? Think of all the cases that are built on what a police officer says – that he found narcotics on the accused and the accused says that the drugs were planted by the police officer. A simple test, and the case is decided."

"Can we even reach the conclusion that it would be unethical *not* to use the technology, if it works with enough certainty?"

"Perhaps. It depends on how you balance all the different aspects. Let's say that somebody didn't want MRI used on him. It would only be unethical not to use it if you think that getting to the truth is more important than the individual's right to control the access to their own brain."

"This suddenly raises the question about the state's right to open up people's brains?"

"Definitely." Greely looks as if he is going to continue down this tangent, but then he puts down his soda can and remarks that it would be wonderfully simple if someone got the ball rolling by voluntarily asking for an MRI test in a concrete case. And yet.

"You can, of course, still decide to deny the request on the grounds that the test is too powerful a piece of evidence."

"I don't quite understand that."

"You see, four of the Supreme Court's nine members have already signed onto a decision that says that MRI lie detectors shouldn't be used, even if they are reliable, simply because they would remove too much power from the jury. I think there will be more reactions like that. Even if a volunteer turns up, there is the possibility of keeping it out, because the jury or the judge would be too impressed by the technology itself. Too much weight would be put on the technology and thereby create a prejudice in the case."

Nevertheless, Greely's guess is that courts will ultimately allow MRI.

"But then questions come up in connection with witnesses," he says, throwing his large body forward in a quick movement. "Can we force them to take a test to determine whether they are testifying correctly? Or if they provide testimony that is supported by their own test, can the opposing party force them to take a test as to *their* claims?"

The lawyer comes out in the professor, and you can see how he enjoys the inner chess game of direct and cross-examination.

"To get back to what you said about the state's right to open up people's brains," he says, "you can ask whether a warrant to search a suspect's house or office could also be used to search his brain."

I play along. "In that case," I say, "where does our right to privacy stand in relation to society's right to find out what we know and whether we're telling the truth?"

"Yes, and ultimately, perhaps, what we are thinking and feeling," adds Greely, smiling broadly, before he takes the conversation in another direction without answering.

"You could ask about the extent to which this is even new territory. We're trying to read people's minds all the time. We're social animals and are always reading other people's intentions and knowledge. There are two things different about MRI technology. First, it is more effective and, second, we're no longer looking at indirect signs but at the brain itself. The organ that generates our thoughts, knowledge and self."

He looks up at the ceiling at an angle. "I don't know whether it really makes a difference. You can try to draw a sharp line at the cranium and say everything inside this bone is forbidden territory, while everything else can be read freely."

This is a subject Greely discusses again and again with his colleague, bioethicist Paul Root Wolpe at Emory University.

"He thinks it is absolutely crucial that we can now look inside people, but as for me I'm not sure it makes the case morally or ethically different."

"Is it time to start discussing cognitive liberty?" I ask quickly. I know that Greely is also familiar with the Center for Cognitive Liberty, where a small group of volunteer lawyers and sociologists are on the cutting edge of the neurorevolution. For several years, they have been talking about how neuroscience opens the door to a society in which the venerable but somewhat airy notion of freedom of thought should suddenly be taken quite literally.

"Cognitive liberty covers a lot. Up at the centre, it's all about the individual's right to change his mental status with drugs, but I don't really know how acutely important that case is."

"But the right to a private mental life?"

"That is certainly important to confront. We have always discussed the right to keep our own ideas and opinions to ourselves, but the prospect of an effective means for reading mental content makes it pressing to have a societal debate. It is clear that developments in neuroscience will increasingly improve the possibilities for linking electrical activity and blood flow in the brain to particular mental states."

Greely suddenly sticks a little finger into the air. It looks like a plump sausage, when he wobbles it.

"Today, we can correlate the movement of a finger with activity in the motor cortex, and we can attribute sight, hearing and speech to other areas of the cerebral cortex. And we may be able to correlate more subtle states."

"Which could be what?"

"Let me just say that all this reminds me of my work with genetics. There are genes that have a powerful effect on people – the gene for Huntington's disease, for example. Then, there are other genetic links that are much weaker. I think we'll see the same thing with mental states. The movement of my finger here is very easy to spot in the brain, but my desire for another diet cola undoubtedly provides a somewhat weaker correlation. And given the fact that we are more interested in questions such as whether you are a liberal, Muslim or Christian, or planning a terrorist attack, this will probably be difficult. But it is an empirical question, and we have to wait and see."

Hank Greely likes to assign categories to the interesting legal questions.

"One is mind-reading. It can be lie detection or measuring the extent to which someone feels pain and suffering and is, therefore, entitled to compensation after an accident or something. It can also have to do with the extent to which someone recognizes a crime scene, and there is reason to believe you can document that using brain fingerprinting."

Beyond mind-reading, there is prediction.

"Will you get Alzheimer's? Is an apparently normal teenager developing schizophrenia? Does a man have a predisposition to sexual deviancy, or does he have a violent personality?"

"Or psychopathic tendencies," it pops out of me. "That might be diagnosed early with a quick brain scan."

"I agree. There are studies of psychopaths, and there will undoubtedly come more. But here I have to go back to genetics again. They have discovered that certain things require subtle interactions between genetics and environment."

I know his example as one everybody pulls out; a study from New Zealand showing that a variant of the so-called MAO-A gene, which encodes an enzyme in the brain, may increase the risk of violent behavior. However, this only happens if the carrier has also had a miserable upbringing with neglect and stress. A prime example that it's not a question of heredity *or* environment but heredity *and* environment working in tandem.

"Parallel to this, I think that things like Alzheimer's will prove to be easier to predict, while violence is certainly more difficult and more uncertain and influenced by more factors. At the same time, I'm convinced that measurement of the brain will prove to give far better predictions than a lot of genetics, because it is an activity that is going on in the person here and now."

But if brain scanning turns out to be more predictive, isn't it important to take a position on what we do with the knowledge? For example, in relation to violent people or sexual deviants?

"Yes, yes," says Greely, who is then quiet for a moment. "People have a tendency to see it as an either-or. Either we lock them up or we leave them be. But you could also begin to see possibilities for therapy or counseling, maybe even medication. And you could warn neighbors or decide that particularly risky people should wear some sort of electronic device, so they can be tracked, as the case may be. Or," Greely says with a raffish grin, "you could tattoo a big red R for rapist on their forehead."

In real life, he believes, new alternatives will arise – not by politicians deciding to force them through but by someone volunteering. Asking to get a pill or an electronic bracelet instead of being stuck in jail.

"In the US, it's already the case that certain states give sex offenders the opportunity to avoid jail if they agree to be castrated – medically or surgically."

"*Surgically*?" I'm a little shocked. "Is that something that goes on down South?"

"How could you guess? It's in Texas."

Greely believes, however, that the prediction of disease will be well on its way long before people start tackling sex and violence.

"Here, I wouldn't underestimate the effect of legislation with respect to discrimination. Can you be insured if you are likely to get Alzheimer's, or can you become a police officer if you have a high probability of developing schizophrenia?"

"Can you get into the military," I ask, thinking of Colin Camerer.

"Yes, or just getting into a good university. There are also politicians – should they be forced to reveal their neurological prospects? Shouldn't we as voters know if this guy, who might lead the country, in all probability will begin to show signs of Alzheimer's in three years?"

I observe that Ronald Reagan, according to various observers, had major symptoms of Alzheimer's during much of his presidency. Greely nods heavily and looks up toward the ceiling.

"The possibility of being able to predict criminality and sexual pathology sounds more sexy and titillating to people than diagnostic tests for disease. But they have little chance of becoming reality, so we need to watch out that we don't get distracted by them."

And what about neuroeconomists? I say. They think that the way your brain behaves in investment situations can be a tool for the big funds and investment banks for employing people.

"That's possible. On the whole, it's interesting that we know more and more about how people make choices, right? For me, it raises concerns about manipulation. If people know, for example, that you are an easy target for sentimental appeals, couldn't Google target its commercials to you accordingly?"

"That's possible," I reply and ask the next logical question: where does good, old free will fit in this whole neurorevolution?

"Ah, that's something people love to talk about, responsibility and free will," says Greely. "And I don't much care for it. Philosophers especially are drawn to this topic like moths to a flame. Apparently, they don't have the free will to avoid it. You just don't get much more out of it than endless arguments about whether, if neuroscience proves that there is no free will, we should even have a legal system. Of course, we should," the lawyer says stridently.

"We will act *as if* we have free will, no matter what a bunch of researchers might say about it."

"If I can just break in here. It does matter if, for example, you can show that a psychopath doesn't do anything he believes or feels is wrong," I say, making reference to Marc Hauser in Boston. "Then he doesn't have guilt in the usual sense, and what do we do then?"

"I guarantee that we will do *something*," retorts Greely impassively. "I think we will continue judging and intervening with respect to people who break established rules whether we believe they are mentally competent or not. Today, we see that any defense based on insanity or mental incompetence ends with the accused being committed to an institution, often for an indeterminate period. They don't go free, right?"

"God forbid. But if the idea of guilt and who is guilty changes on the basis of research," I respond, "wouldn't it put pressure on the public to give a lot of lawbreakers a different treatment than prison?"

"I doubt it," says Greely, without blinking. "It is, of course, an open question, but I think we will continue to feel that people who break rules should be punished. If neuroscience can show us better *ways* to deal with them than those we have today, we might listen. It is a more likely way for the research to influence law – not with respect to the question of guilt but with respect to the practical response."

Nevertheless, he lets himself speculate. It may well be that, in a few years, we will see a change in whom we recognize as insane and in the possibility for testing whether someone is faking. Maybe brain scanning can show whether a person has hallucinations or not. And maybe scanners can reveal whether conditions in a person's anterior cingulate cortex determine whether the person in question has impulse control.

"But mental competence is not just an issue with respect to criminality. It also has to do with the extent to which people are competent to take care of themselves and especially their money. I can imagine a society in which old people have to undergo a brain scan before they can write a will."

"Can the neurorevolution lead to a new evaluation of the balance between individual freedom and concern for the community? If we are able to predict or ascertain that some individuals are a danger to the many – won't it suddenly become more important to protect us than them?"

"It's possible."

"Yes, but aren't the times primed for exactly this sort of thing?"

"The times are always primed for this sort of thing. There have always been threats and wars and invasion, and what have you. Our notion of individual freedom is under pressure right now because of terrorism, but we've seen this before – here in this country you can point to the internment of Japanese-Americans during the Second World War and McCarthyism's witch hunt for Communists in the 1950s. All pure fear. Neuroscience can give us better, more precise tools to be used in times of crisis. I'm all for that, but I don't think they can in themselves shift the balance between the individual and the community in any permanent way."

When Greely continues after his last sip of diet cola, it is with a lighter voice.

"Or if they do, maybe it will be more in the direction of a focus on the individual. Looking into people's brains has, above all, the potential to show *how* different and individual we all are."

It may be that Hank Greely the lawyer is irritated by armchair philosophers and not much for discussing free will. But after our meeting, this is the one issue that keeps circling in my own head. The concept of free will is the very cornerstone of our legal system, and all our conceptions of guilt are tightly linked to the notion that the guilty person *did it on purpose, from free will.*

Now comes brain research, putting a big, fat question mark on whether the will is indeed free. There are tantalizing experiments like the one coming out of the Max Planck Institute in Leipzig where John-Dylan Haynes and his team scanned a group of people making decisions about whether to push one of two buttons.[47] It turned out the researchers could predict the outcome seconds before the person felt he had made up his mind. What they looked at was activity in brain regions associated with *unconscious* processing. As Haynes said to a reporter from *Science*: "When it comes to decisions we tend to assume they are made by our conscious mind. This is questioned by our current findings."

Indeed it is and most researchers will say that the classical view of "free will" is automatically blown out of the field as soon as we accept the fact that people are exclusively physical beings. You can't talk about the will as a completely independent thing that acts in a sort of mental vacuum, elevated above causal connections. Brains are making the decisions, and brains are physical systems that operate causally – every current state is dependent on the previous state.

But how can we update the metaphysical concept of free will? One idea comes from Patricia Churchland, a neurophilosopher at the University of San Diego, who argues that we need to start talking about self-control. Much less fluffy than free will, self-control is a character trait that obviously varies by different degrees among species and even individual people. And it can be studied. The limitations and conditions that exist for self-control in individual brains and under different circumstances can be explored directly. Researchers can figure out what brain structures are involved in the control of various types of behavior and determine how this control can be weakened or strengthened.

In relation to the legal system, concepts such as guilt and responsibility are directly connected to the idea of control. Seen in this light, brain research should be the stumbling block that initiates a fundamental debate about how the legal system should function. This is a new debate. We don't usually question the foundations of the system but stick to debating how penalties should be adjusted. One week there's a demand for stiffer sentences for pedophiles, the next week it's hit-and-run drivers. But there is an urgent need to discuss *why* we punish and *what* we as a society want punishment to achieve.

Joshua Greene, the moral philosopher turned neuroscientist, is one of the brave few who has begun to touch on the issue both in academic journals and media interviews. Accepting that we *are* our brains and that our wills are not completely free, he argues, should lead the criminal justice system to abandon the idea of retribution. When dealing with criminals, says young Greene, the system should focus narrowly on deterring future harms. In this thinking, the underlying purpose of punishment is not to make the general public feel that "criminals get what they deserve" but rather to minimize crime.

The logical next step is the discussion of whether to use knowledge about brains in crime prevention. While criminality is not a disease, it is *partially* an expression of biological variation – the ability to resist temptation is not purely accidental. And if – or when – it becomes possible to assess how much self-control an individual possesses and, perhaps, in what circumstances the person is especially vulnerable, this could play a role in deciding their future. You can discuss the possibilities of treatment. You can speak of punishing individuals based on assessments of the risk that they might re-offend. You might even be able to get into a discussion about the extent to which you can identify especially vulnerable people *before* they ever commit a crime and possibly intervene preemptively.

These are explosive issues. They are also issues that can't be ignored, as ever-more intimate details of human nature are exposed by the scanners.

9

FREE US FROM OURSELVES

WHAT HAPPENS TO THE HUMAN MIND WHEN IT COMES TO KNOW ITSELF ABSOLUTELY?

Tom Wolfe's razor-sharp, provocative question instigated my sojourn into brain research, and persists to the very end. It is this fundamental question that the law professor, the philosophers and scientists I talked to along the way have wrestled with, each in their own way, each through their own professional filter.

Of course, the very idea that we might be able to understand ourselves *absolutely* must be taken with a grain of salt. Many scientists would say the brain is quite simply too complex to give up all its secrets, and so far they are right. But the completeness of the endeavor is not crucial, for no one doubts that we can achieve a very solid understanding of this eternally fascinating organ. It's still early, and researchers are merely scratching the surface when they link mental phenomena with a change of blood flow here and some electrical activity there. Yet there have been massive advances over just a few decades, and it is not unreasonable to expect a quantum leap or two over the next ten to twenty years. In forty or fifty years, who knows? In any case, neuroscience attracts so much money, such intense interest, and so many great minds that breakthroughs are inevitable.

So what should we expect between now and then?

Undoubtedly, great things. A better understanding of the brain will undoubtedly – and visibly – lead to a better understanding and treatment of conditions that today are more or less mysterious: Alzheimer's, schizophrenia and depression, to mention only the most obvious. This is progress that is tangible even now. But developments in the future will undoubtedly bring new technologies that will leave us gawking. Even today, electronics connected directly to nerve tissue can make the deaf hear and very soon will enable the blind to see. A bit further down the road, it is not unlikely that intelligent chips implanted in just the right place will allow the lame to stand and walk. And the predictions are that computer technology will at some point be able to upgrade normal neurological functions by connecting the brain to external computers, allowing us to export and import information directly. At the same time, information *about* the brain will be used in a great many areas, of which marketing, lie detection, and management are just a few. In just half a century, when we look back over our shoulders, things will be happening that we can't imagine today.

The technological progress will be amazing, but the technology itself will not be the most interesting thing. The most meaningful and sweeping change will take place – as Tom Wolfe realized long ago – in the mental domain, in our *minds*. A parallel can be drawn to the DNA revolution, which has just celebrated its silver jubilee. Over the course of the last five decades, genetic information and gene technology have gradually permeated virtually every sector of society. Genes are decoded, and the information is used in countless ways from food production to manufacturing to the health industry. At the same time, the *idea* of gene technology has percolated into the general mindset, where it remains an obvious option and – with a bit of good will – an entirely natural thing. For the first time, human beings have become the "masters of creation." Gene technology has made us capable of engineering living organisms and thus effectively capable of moving beyond the limits of evolution. And with these possibilities, the genetic revolution is radically reorganizing our psychological landscape. Our view of life and our relationship to nature is changing. It is becoming clear that life is not a question of fixed forms but, ultimately, a stream of information – information that can be combined freely – and that nature is something we can construct on an equal footing with evolution.

What will define the neurorevolution? As I see it, we are looking at a liberation. The general message wrapped up in the countless research results is a message of freedom. That might sound overblown, but it becomes obvious when you think about what all this brain research is fundamentally *doing*. It is providing us with unprecedented insight into ourselves. The scanners and artful experiments are pushing us past guesswork, assumptions and vague notions in order to reveal exactly what is hiding deep within this formidable creature *Homo sapiens*. By exposing human nature, neuroscience makes us able to transcend and rise above it.

This radical self-knowledge arms us with insight that allows us to reflect on our own essence, on our feelings, motives, attitudes and actions at a higher level than we have ever been able to before. We are no longer limited to experiencing things subjectively, but are in a position to understand intellectually what sort of processes are creating our subjective experience. In particular, we're about to have our eyes opened to what emotions really are – quick assessments that take place in the unconscious and which need not be especially "true". In other words, signals that we don't necessarily need to obey.

Think of the trendy neuroeconomists casting a bright light on how we make decisions. When we *understand* why we react the way we do in a choice situation, and understand how it happens that we have certain specific feelings and needs, it suddenly becomes easier to look at them objectively. Take, for example, the punishment we are so willing to inflict upon our fellow human beings. Wouldn't it slake our desire for vengeance a little, if we knew that the need to sanction and correct is a simple hereditary reaction that once served as a survival mechanism? Parents and educators, colleagues and even politicians preach tolerance on the grounds that it will make everyday life run much more smoothly, but wouldn't the argument be more convincing if you could see for yourself exactly where and why intolerance arises? And the same is true of all other emotions and intuitions.

Of course, this means that the knowledge we glean from brain research must be transformed into collective general knowledge, and integrated the same way Freudian thinking is today. Without even thinking about it, and for a lot of people without even realizing where it comes from, we analyze the world around us and other people's motives, utterances and actions with themes from Freud in the back of our minds; the traumas of childhood that form and influence the adult, sex as an omnipresent motivator, and so on. This is a reflex, it is a part of culture and practically impossible to ignore.

One of the neurosociety's most sweeping changes will be in our vision of humanity, our conception of who we are. It reflects a point that happiness researcher Richard Davidson made – namely, that the self is a far more mutable specimen than we have traditionally realized. "A more fluid self," he said. And this image is becoming more and more inevitable, as the conception of an incorporeal soul disappears and the self is inexorably located in the brain and ascribed to an interplay of cells and chemical processes. For when you begin to look for the self, you can't really find it. Researchers, who have only just begun to tackle this huge question, simply cannot localize any particular place in the brain where the feeling of self belongs or can be formed. They theorize that a number of networks must exist that create aspects of what we consider a self. This is an interesting break with traditional thinking. For how do we typically react, when faced with someone who slowly changes character because of dementia or people who wake up after a cerebral hemorrhage and have a hard time finding their way back to their lives?

"They are no longer themselves," we say. The fact of the matter is rather that there *isn't* any self in that sense. We have no fixed essence of identity.

The German philosopher Thomas Metzinger of the University of Mainz expresses it quite beautifully with his "phenomenal self model," which he describes in his book *Being No One*. Whereas traditionally we were each "someone" with a conception of an unassailable core – the "I," we must now acknowledge that this self is not a substance but representations of the information that is processed in the brain. The phenomenon of self is like the light in a light bulb – a transitory dynamic phenomenon that arises as the discharge of certain physical processes. In a way, we *are* these processes. There is no tiny homunculus in the head and the *feeling* that there is this "self" is just a feature built into the system. As Metzinger writes: "Certain organisms are in possession of conscious self models, but such models in no way constitute a *self* – they are simply complex states in the brain."

One state, one self, another state, another self, you might say. And a recognition of the fluid self sets the stage for a recognition that life is not so much about finding yourself but choosing yourself or molding yourself in the shape you want to be. Again, we're talking about a message of freedom and anti-determinism: your personal biology is not a prison but a lump of playdough you can help form.

Of course, self-development is nothing new. Freud and the psychological revolution drove home the point that we *can* change our psyche, and the notion of realizing ourselves has been a powerful trend in late modern Western society. In the future, self-development will simply be far more specific and targeted. The acceptance of the fluid self will take place as a matter of course, particularly because it builds on modern thinking in which identity is something to be played with. In the latter half of the twentieth century, we learned that gender was something you could explore and fool around with in your dress and general *modus vivendi*, and even sexuality has become a more or less open field, where the boundaries for what is normal and accepted are constantly expanding. Even more recently, we've discovered virtual reality – that boundless cyber-universe – a whole new territory in which we can express ourselves with all the usual rules suspended. Here, identity can be chosen freely, independent of the physical world.

The neurotechnology of the future will likewise provide the means for transforming and forming the physical self – be it through various cognitive techniques, targeted drugs, or electronic implants. And we are going to use these tools in our private projects without fear and trembling, because our individual self will simply be a broad range of *possible* selves.

Does this sound seductive? A little too rosy? Of course, you might object that we're talking about a positive interpretation, a sort of best-case scenario. The truth is undoubtedly that, at first, not everyone will rejoice at all this self-insight provided by the neurorevolution. The thing is that, with the acceptance of the fluid, strictly physiological-chemical self, we will be dragged into a radically and brutally naturalistic interpretation of humankind with which we have never lived at any point in our history. To date, the predominant model has been metaphysics. People are something *more* than we can measure. Removing this "more" will be an existential shock that must be absorbed by the culture; which will, of course, resist.

Here, too, we can compare the progress of gene technology, which did not arrive without a fuss. Throughout much of the Western world, the past fifteen years have seen loud and bitter protests over genetically modified crops and "Franken food," and there is an ongoing heated debate about cloning and embryonic stem cells. All this commotion is a sign that we find ourselves in a transitional period. People feel insecure and, for many, there is a certain instinctive disgust at the whole new mind set.

The neurorevolution, too, will provoke disgust. The simple burgeoning of opportunities for influencing one's own life will increase frustrations for many – for what should one choose? At the same time, a lot of people will experience emotional stress, because they will feel lost without a metaphysical crutch. One unavoidable consequence of the soul's final farewell is that there is no meaning to life – you can't derive any meaning from above or, for that matter, from biology and the physical world itself. The message is that we each have to create meaning in our own lives and the way we live it. And frankly, there isn't much comfort in that.

I imagine we will see two disparate trends. On one hand, more and more people will absorb and live with the scientific worldview fostered by the results of neuroscience. On the other hand, the same research and its unveiling of human nature will provoke a counter-reaction. For those who cannot bear to say goodbye to metaphysics and a spiritual standpoint, there will be no place to turn except fundamentalist religion – where faith is allowed to monopolize the truth and reject everything else. Between these two extremes, "slight discomfort" will take the middle ground. And it will be manifested as a resistance to specific neurotechnologies.

As it gradually becomes practical to read more and more aspects of our mental habitus, there will be groups of concerned citizens whipping up fears. Should it be legal to invade our brains and reveal our thoughts and feelings to employers, insurance agents or presiding magistrates? And what about children? May their parents force them into a scanner to get an overview of their potential? Doesn't the very existence of these technologies lead directly to some version of *1984*? And don't the fields of neuromarketing and neuromanagement open the gates for manipulating people and, perhaps, the whole of society in the grossest manner? Not to mention the long-term possibilities for changing your self. Should people who can afford to do so be able to use existing technologies to upgrade themselves at some point? And society – should the authorities be able to use new technologies to prevent or treat deviants who are in some way a danger to themselves or others?

To date, these sorts of questions and speculations haven't occupied much space in the fluttering public consciousness, and the little attention they have drawn is limited to a very narrow circle. A few years ago, the word "neuroethics" made its debut at a conference for bioethicists. Today, there is an international society of self-proclaimed neuroethicists who have academic discussions at meetings and in journals. Beyond them, you can find a handful of blogs and homepages, where those interested exchange viewpoints on the latest science and keep stressing to each other that now we really *have* to get a broader social debate going.

Of course, they're right, there is an urgent need for a broader conversation and especially one that involves the political establishment. The same politicians who are talking up the "knowledge society" and urging on new research and technology need to open their eyes to the broader implications of the path they are beating. When gene technology was emerging and making its way into products, politicians were caught napping when the popular reaction hit. They had not anticipated the anxieties the technology would cause and they never got a handle on the situation – and consequently are still hobbling after the debate without any clear standpoint.

They have a chance to get ahead of the curve in the next great upheaval, and they have a responsibility not to let the shock of the new and public fear of technology run away with them. Standing on the threshold of the neurorevolution, we should use our intellect and let our curiosity take us to the heart of these new possibilities. We might wind up in a fruitful discussion of how we can use knowledge about the brain and brain technology *offensively* to achieve what we all want: the good life. At which point, we're back with the point made by Iacoboni and Damasio that all this deeply-fascinating brain research can and should be a way to improve the individual and society as a whole.

What is needed to generate that focus and kick-start the discussion?

When I put this question to sociologist Paul Root Wolpe, who is deeply involved in neuroethics, he answered without blinking. "We need a shocking case – something like Dolly, the cloned sheep, before the public discovers that this is important," he said. Then, he added, almost without missing a beat: "And I'm convinced we'll get one."

Notes

1. *The Descent of Man*, p. 69.
2. Kelemen 2004, *Psychological Science*, vol. 15, no. 5, pp. 295–.
3. Newberg *et al.* 2006, *Psychiatry Research: Neuroimaging*, vol. 148, no. 1, pp. 67–71.
4. Beauregard *et al.* 2006, *Neuroscience Letters*, vol. 405, pp. 186–90.
5. Persinger & Koren 2001, *Perception and Motor Skills*, vol. 92, no. 1, pp. 35–6.
6. Baker-Price & Persinger 2003, *Perception and Motor Skills*, vol. 96, pp. 965–74.
7. Khamsi 2004, *Nature*, published online 9 December.
8. Hauser *et al.* 2007, *Mind and Language*, vol. 22 (1), pp. 1–21.
9. Hauser M.D. *et al.* 2006, *Social Cognitive and Affective Neuroscience Advance Access*, October 20.
10. Greene *et al.* 2001, *Science*, vol. 293, pp. 2105–8.
11. Greene *et al.* 2004, *Neuron*, vol. 44, pp. 389–400.
12. Brosnan *et al.* 2003, *Nature*, vol. 425, pp. 297–9.
13. www.mises.org/story/1893.
14. Knobe *et al.* 2003, *Analysis*, vol. 63, pp. 190–3.
15. Urry *et al.* 2004, *Psychological Science*, vol. 15, no. 6, pp. 367–72.
16. Richard Davidson 2004, *Philosophical Transactions of the Royal Society*, vol. 359, pp. 1395–1411.
17. Lyubomirksy *et al.* 2005, *Review of General Psychology*, vol. 9, no. 2, pp. 111–31.
18. Lyubomirsky & Ross 1997, *Personality and Social Psychology*, vol. 73, pp. 1141–57.

19. Boehn & Lyubomirsky 2007, in S.J. Lopez (Ed.) *Handbook of Positive Psychology*, Oxford University Press.
20. Lykken & Tellegen 1996, *Psychological Science*, Vol. 7, no. 3, pp. 186–9.
21. Urry *et al.* 2004, *Psychological Science*, Vol. 15, no. 6, pp. 367–72.
22. Buss *et al.* 2003, *Behavioural Neuroscience*, vol. 117, no. 1, pp. 11–20.
23. Lutz *et al.* 2004, *PNAS*, vol. 101, pp. 16369–73.
24. Davidson *et al.* 2003, *Psychosomatic Medicine*, vol. 66, pp. 564–70.
25. Urry *et al.* 2006, *Journal of Neuroscience*, vol. 26, pp. 4415–25.
26. Phelps *et al.* 2000, *Journal of Cognitive Neuroscience*, Vol. 12, no. 5, pp. 729–38.
27. Cunningham *et al.* 2004, *Psychological Science*, vol. 15, no. 12, pp. 806–13.
28. Wilson *et al.* 2004, *Nature Neuroscience*, vol. 7, no. 7, pp. 701–2.
29. Iacoboni, personal communication.
30. Keysers *et al.* 2004, *Neuron*, vol. 42, no. 2, pp. 335–46.
31. Iacoboni *et al.* 2005, *PloS Biology*, vol. 3, no. 3, pp. 529–35.
32. Burman *et al.* 2008, *Neuropsychologia*, vol. 46, 5, pp. 1349–62.
33. Sanfey *et al.* 2003, *Science*, vol. 300, pp. 1755–8.
34. McClure *et al.* 2004, *Science*, vol. 306, pp. 503–7.
35. Berns *et al.* 2006, *Science*, vol. 312, pp. 754–8.
36. Dominique *et al.* 2004, *Science*, vol. 305, pp. 1254–8.
37. Kosfeld *et al.* 2005, *Nature*, vol. 435, pp. 673–6.
38. King-Casas *et al.* 2005, *Science*, vol. 308, no. 5718, pp. 78–83.
39. Lo & Repin 2002, *Journal of Cognitive Neuroscience*, vol. 14, pp 323–39.
40. Kuhnen & Knutson 2006, *Neuron*, vol. 47, pp. 763–70.
41. McClure *et al.* 2004, *Neuron*, vol. 44, pp. 379–87.
42. Langleben *et al.* 2002, *Neuroimage*, vol. 15, pp. 727–32.
43. Ganis *et al.* 2003, *Cerebral Cortex*, vol. 13, no. 8, pp. 830–6.

44. Vendemia *et al.* 2005, *American Journal of Psychology*, vol. 118, no. 3, pp. 413–29.
45. Farwell and Smith 2001, *Journal of Forensic Sciences*, vol. 46, no. 1, pp. 1–9.
46. Kozel *et al.* 2005, *Biological Psychiatry*, vol. 58, no. 8, pp. 605–13.
47. Soon *et al.* 2008, *Nature Neuroscience*, vol.11, pp. 543–5.

Index